Carbon Functional Nanomaterials, Graphene and Related 2D-Layered Systems

MATERIALS RESEARCH SOCIETY
SYMPOSIUM PROCEEDINGS VOLUME 1549

Carbon Functional Nanomaterials, Graphene and Related 2D-Layered Systems

Symposia held April 1–5, 2013, San Francisco, California U.S.A.

EDITORS

Mauricio Terrones
The Pennsylvania State University
University Park, Pennsylvania, U.S.A.

Swastik Kar
Northeastern University
Boston, Massachusetts, U.S.A.

Ken Haenen
Hasselt University and IMEC vzw
Diepenbeek, Belgium

Pulickel M. Ajayan
Rice University
Houston, Texas, U.S.A.

Jose Antonio Garrido
Technische Universitaet Muenchen
Garching, Germany

Anupama Kaul
California Institute of Technology
Pasadena, California, U.S.A.

Cheol Jin Lee
Korea University
Seoul, S. Korea

Joshua A. Robinson
The Pennsylvania State University
University Park, Pennsylvania, U.S.A.

Jeremy T. Robinson
Naval Research Laboratory
Washington, D.C., U.S.A.

Ian D. Sharp
Lawrence Berkeley National Laboratory
Berkeley, California, U.S.A.

Saikat Talapatra
Southern Illinois University Carbondale
Carbondale, Illinois, U.S.A.

Reshef Tenne
Weizmann Institute of Science
Rehovot, Israel, U.S.A.

ASSOCIATE EDITORS

Ana Laura Elias
The Pennsylvania State University
University Park, Pennsylvania, U.S.A.

Mandar Paranjape
KLA-Tencor Corporation
Milpitas, California, U.S.A.

Neerav Kharche
Brookhaven National Laboratory
Upton, New York, U.S.A.

Materials Research Society
Warrendale, Pennsylvania

CAMBRIDGE
UNIVERSITY PRESS

University Printing House, Cambridge CB2 8BS, United Kingdom

One Liberty Plaza, 20th Floor, New York, NY 10006, USA

477 Williamstown Road, Port Melbourne, VIC 3207, Australia

314-321, 3rd Floor, Plot 3, Splendor Forum, Jasola District Centre, New Delhi - 110025, India

79 Anson Road, #06-04/06, Singapore 079906

Cambridge University Press is part of the University of Cambridge.

It furthers the University's mission by disseminating knowledge in the pursuit of education, learning and research at the highest international levels of excellence.

www.cambridge.org
Information on this title: www.cambridge.org/9781605115269

Materials Research Society
506 Keystone Drive, Warrendale, PA 15086
http://www.mrs.org

© Materials Research Society 2013

First published 2013

CODEN: MRSPDH

A catalogue record for this publication is available from the British Library

ISBN 978-1-605-11526-9 Hardback

CONTENTS

*Invited Paper

OTHER 2D-LAYERED MATERIALS

CARBON NANOTUBES

CARBON NANOMATERIALS

PREFACE

For almost two decades now, the carbon nanomaterial (CNM) system has persistently provided researchers the opportunity for spectacular new discoveries, significant advances in fundamental and applied science, and the development of disruptive technologies and applications. The rich allotropicity of carbon bonding can explain the broad use of carbon-based materials such as carbon nanotubes (CNT), diamond, fullerenes, and more recently graphene. Today's research community continues to discover and harness new low-dimensional carbon allotropes, perhaps at a historically unprecedented rate. In this context, carbon nanotubes, nanodiamond, and graphene, have become versatile platforms for new materials properties and device architectures, and are finding their way into nearly every facet of the research world, including conductive polymers, transparent electrodes, chemical sensors, high-frequency devices, optoelectronic sensors, alternative energy, and bio-inspired systems, to name a few. At the same time, researchers from diverse disciplines are pushing the frontiers of these materials by developing innovative arrays of ribbon, hybrid, functionalized, doped, and hetero structures often resulting in dramatically new scientific and engineering directions.

The significance of CNMs is demonstrated by the daily stream of new research publications – many of which are in journals of highest impact factors – addressing a broad range of experimental and theoretical materials-related topics including synthesis, characterization, integration, and devices. Interdisciplinary topics related to the materials science, chemistry, physics, mechanics, and engineering of CNMs such as graphene, carbon nanotubes, nanoribbons, nanodiamond, graphene oxide, graphane, fluorographene, graphene composites (and many others) were the focus of these symposia, with a long-term outlook on applications of these materials.

Within these CNMs, the isolation of graphene has been a turning point which has resulted in the emergence of a new research area namely "atomically-thick 2-Dimensional systems", in which monolayers of layered materials such as BN, BCN, MoS_2, WS_2, etc., are now being isolated and studied. In contrast to the graphene gapless density of states, other 2D systems could possess well-defined and tunable electronic gaps, thus offering numerous potential applications in nanoelectronics and optoelectronics, such as sensors, logic devices, high energy rechargeable batteries, photodiodes, etc. However, there are numerous challenges that need to be overcome regarding the synthesis of mono-, bi-. tri-layers as well as their optical/electronic characterization. All these exciting developments and others, are also covered in the following contributions and we hope the reader find them relevant and useful when carrying out fundamental research in the areas of carbon nanoscience and 2D systems.

Mauricio Terrones
Swastik Kar
Ken Haenen
Pulickel M. Ajayan
Jose Antonio Garrido
Anupama Kaul
Cheol Jin Lee
Joshua A. Robinson
Jeremy T. Robinson
Ian D. Sharp
Saikat Talapatra
Reshef Tenne
Ana Laura Elias
Mandar Paranjape
Neerav Kharche

November 2013

MATERIALS RESEARCH SOCIETY SYMPOSIUM PROCEEDINGS

MATERIALS RESEARCH SOCIETY SYMPOSIUM PROCEEDINGS

Prior Materials Research Symposium Proceedings available by contacting Materials Research Society

Graphene and Beyond Graphene

Mater. Res. Soc. Symp. Proc. Vol. 1549 © 2013 Materials Research Society
DOI: 10.1557/opl.2013.858

Layered Nanostructures – Electronic and Mechanical Properties

Gotthard Seifert[1], Tommy Lorenz[1] and Jan-Ole Joswig[1]
[1]Physical Chemistry, Technical University Dresden, 01062 Dresden, Germany

ABSTRACT

In addition to graphene, 2D transition-metal chalcogenides as, e.g., MoS_2 and WS_2 nanostructures are promising materials for applications in electronics and mechanical engineering. Though the structure of these materials causes a highly inert surface with a low defect concentration, defects and edge effects can strongly influence the properties of these nanostructured materials. Therefore, a basic understanding of the interplay between electronic and mechanical properties and the influence of defects, edge states and doping is needed. We demonstrate on the basis of atomistic quantum-chemical simulations of a circular MoS_2 platelet, how the mechanical deformation can vary the electronic properties and other device characteristics of such a system.

INTRODUCTION

The electronic properties of semiconducting transition metal dichalcogenides MX_2 (M: Mo, W; X: S, Se) with hexagonal 2H crystal structure have been studied already in the 1960s, e.g. in the pioneering work of Fivaz and Mooser [1]. Layers of dichalcogenides are characterized by a metal sheet sandwiched between two sheets of chalcogenide atoms; a typical example is MoS_2. The molybdenum atoms are six-fold coordinated in this environment, and in the most common phase the sulfur atoms form a prismatic coordination around each molybdenum atom.

Besides graphene, also other layered materials can be exfoliated down to single sheets. This has been shown for a number transition metal dichalcogenides some time ago [2] and revealed also by Novoselov et al. [3]. As single layers, the dichalcogenides of molybdenum and tungsten are semiconducting. Galli et al. [4] showed that – going from the bulk 2H structure to a single MoS_2 layer (1L-MoS_2) – the system becomes a direct-gap semiconductor. Being semiconducting, single-layer transition metal dichalcogenides are much more attractive candidates for a realization of field-effect transistors than graphene, in which the missing gap creates serious problems for its application in corresponding devices. Consequently, the fabrication of a field-effect transistor based on single-layered MoS_2 has been demonstrated already [5].

Another potential field of applications for MoS_2 is in flexible electronic devices. For such applications the interplay between mechanical stress and the electronic properties has to be understood well. Measurements on MoS_2 and WS_2 nanotubes showed exceptional mechanical properties [6]. Bertolazzi et al. [7] reported on the measurement of the in-plane elastic modulus and breaking strength of single- and bi-layered MoS_2. They used nano-indentation with an atomic force microscope for the measurement of the nanomechanical properties of an ultrathin MoS_2 layer suspended over circular holes.

Here, we demonstrate on the basis of atomistic quantum-chemical simulations of an MoS_2 platelet, how the mechanical deformation can vary the electronic properties. Therefore, we simulate the nano-indentation experiment as described in the theory section. The simulation experiments were performed in close collaboration with our experimental partners, in order to accompany and explain the experimental findings.

THEORY

The calculations were performed using a density-functional based tight-binding (DFTB) method [8,9,10] as implemented in the deMon software package [11]. In the DFTB approach, the single-particle Kohn-Sham eigenfunctions are expanded in a set of localized atom-centered basis functions, which are determined by self-consistent density-functional calculations on the isolated atoms employing a large set of Slater-type basis functions. The effective one-electron potential in the Kohn-Sham Hamiltonian is approximated as a superposition of atomic potentials, and only one- and two-center integrals are calculated to set up the Hamilton matrix. A minimal valence basis set is used.

The studied system is a circular cutout of a single MoS_2 sheet with a diameter of 60 Å consisting of 912 atoms and a tip consisting of 65 molybdenum atoms as depicted in Figure 1. Perpendicular to the sheet, the molybdenum tip is used to simulate the indentation experiment. The DFTB method including a dispersion correction was used to perform a series subsequent Born-Oppenheimer molecular-dynamics (MD) simulations on this system, whereby the edge atoms of the circular sheet (223 atoms located 4 Å or closer to the edge) were not allowed to move. Each part of the series was started by moving the tip towards the sheet by 0.1 Å compared to the simulation before and fixing it at that position. Then the system was equilibrated at 300 K for 1 ps with an MD timestep of 2 fs, in order to simulate the new adjustment of the sheet to the indentation. The equilibrium criterion was $\langle T \rangle - T = \pm 10\,\text{K}$.

Figure 1: Simulation setup for the nano-indentation simulation experiment with a circular MoS_2 platelet and a molybdenum tip. Left side: initial setup; right side: after indentation. The circular cutout of a single MoS_2 sheet consists of 912 atoms, of which the edge atoms are fixed. The tip consists of 65 molybdenum atoms, which are moved towards the layer successively, but fixed during each MD simulation.

DISCUSSION

Mechanical Properties

The described procedure for the simulation of a nano-indentation experiment by means of molecular dynamics and an electronic-structure method gives access to a number of properties, e.g. the total energy and the electronic states at each step. In Figure 2, the total energy as a function of deflection, i.e. depth of penetration, is monitored. As expected, we find an increase in energy with increasing deflection. The curve progression can be fitted by

$$E = a_0 + a_1 \cdot \delta^2 + a_2 \cdot \delta^4. \tag{1}$$

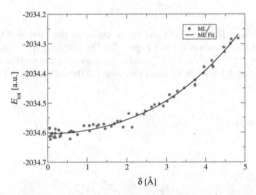

Figure 2: Calculated total energy as a function of the deflection δ resulting from the described simulation setup. Each point results from averaging the total energy over the simulation period of the respective deflection. The fit has been performed according to Equation (1).

From the total energy as function of deflection, we calculated the force F as the first derivative of the energy given in Equation (1) with respect to the deflection δ:

$$F = \frac{dE}{d\delta} = 2a_1 \cdot \delta + 4a_2 \cdot \delta^3. \tag{2}$$

For a rigidly clamped circular plate with radius r and thickness h ($r \Box h$ and $\delta \Box h$), to which no residual stress is applied, a relation between the applied force F at the center of the plate and the deflection δ can be derived [12,13] as

$$F = E^{2D} \frac{q^3 \delta^3}{r^2}, \tag{3}$$

where r indicates the radius of the plate, E^{2D} its surface-based Young's modulus, and q is connected to its Poisson ratio v resulting from

$$q = \frac{1}{1.05 - 0.15v - 0.16v^2} .$$ (4)

Equation (4) is valid for large deflections.

For a negligible bending stiffness and small stress the force-deflection behavior follows a linear progress [13] as

$$F = a \cdot \delta .$$ (5)

Here, the parameter a is connected to the stress [7] by

$$a = \sigma_0^{2D} \cdot \pi .$$ (6)

Therefore, we used the sum of Equations (5) and (6) to calculate the force and the integrated function for the energy fit (cf. Equation (1) and Figure 2). The resulting force as a function of deflection is plotted in Figure 3. For small deflections (negligible bending stiffness and small load), a nearly pure linear dependence can be observed, whereas the cubic term in Equation (2) is dominant for large deflections.

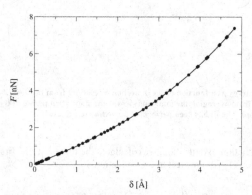

Figure 3: Calculated force as a function of deflection δ. The force is calculated as the first derivative of the total energy (Figure 1) with respect to the deflection according to Equation (2).

From the parameter a_2 in Equation (2), the surface-based Young's modulus E^{2D} can be obtained as $E^{2D} = 4a_2 r^2 / q^3$. We calculated a value of 241 GPa, which is in good agreement with the experimentally observed value of 270 ± 100 GPa [7] and the measured Young's modulus of bulk MoS$_2$ (209 GPa) [14]. Additionally, the maximum stress can be received as

$$\sigma_{max} = \frac{1}{h} \cdot \sqrt{\frac{F^{max} E^{2D}}{4\pi r_{tip}}} \qquad (7)$$

where F^{max} is the maximum force, r_{tip} the radius of the tip and h is the thickness of the layer. From our simulations, we resulted in a value of 22.14 GPa. This value is approximately 10 % of the Young's modulus calculated before, and it agrees very well with the experimental measurements $(22 \pm 4$ GPa) [7].

Electronic Properties

Figure 4 displays the partial electronic density of states (DOS) for different deflections. For the partial DOS, all atoms of the central part of the circular setup, i.e. lying within a radius of 8 Å around the center, have been considered, in order to avoid influence of edge states and to show the contributions of the atoms that are located in the part that is mostly disturbed by the indentation procedure. It can be seen that the main features of the partial density of states are only weakly influenced by small deflections. However, larger indentation depths lead to an evolution of new features in the partial density of states. The new evolving unoccupied states reduce the original band gap.

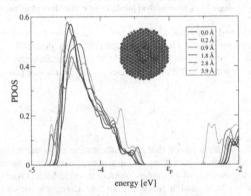

Figure 4: Partial electronic density of states for different deflections. For the partial DOS, all atoms within a radius of 8 Å around the center have been considered.

For comparison, we show in Figure 5 the band gap of an infinite single MoS$_2$ layer as a function of biaxial strain. The diagram shows a strong influence of the biaxial strain on the gap. The maximum deflection in Figure 4 is 3.9 Å. This results in a local strain of the central molybdenum atoms of up to 3% and a reduction of the bang gap by approximately 0.4 eV. Although the strain in our simulated experiment is locally different, the result can be compared to the infinite layer (Figure 5), in which the reduction of the band gap is approximately 0.6 eV at applying 3% biaxial strain. However, we point out again that the mechanical stress in the infinite case is similar for all molybdenum atoms, whereas it is locally different in the simulated experiment.

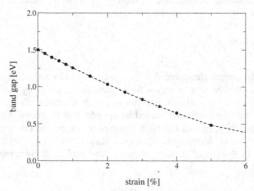

Figure 5: Band gap of an MoS$_2$ layer as function of biaxial strain. The values result from static DFTB calculations of a single MoS$_2$ layer with applied periodic boundary conditions.

The presented simulated nano-indentation experiments showed that fixed single MoS$_2$ layers show a robust behavior of the electronic properties upon deflection. The electronic properties in terms of density of states and band gap do not change abruptly, but vary slightly with increasing deflection. However, in the extreme cases, e.g. at the rupture point, this statement will not be valid.

CONCLUSIONS

In the present study, we have shown that nano-indentation experiments can be simulated very well with molecular-dynamics approaches using a density-functional based tight-binding method. The simulation setup consists of a circular cutout of a single MoS$_2$ layer with a diameter of 60 Å, which was partly fixed, and a molybdenum tip. Using the described simulation procedure, we were able to study the influence of the deflection on certain mechanical and electronic properties.

From the monitored total energy of the system, we were able to obtain the force acting on the layer as a function of deflection. Both, total energy and force are increasing with increasing deflection, whereby the force shows a linear dependence on the deflection for small deflections. For larger deflections the cubic term in Equation (2) is dominant. The resulting Young's modulus and maximum stress were in very good agreement with experimental findings.

The partial electronic density of states of the most central part of the layer showed that the influence of the deflection initially is only weak. However, larger indentation depths lead to an evolution of new features in the partial density of states and reduce the band gap. Compared to a single periodic layer, the change of the band gap upon indentation is smaller than for the infinite system. However, the mechanical stress is similar for all molybdenum atoms in the infinite case, whereas it is locally different in the simulated experiment. The gap has been calculated from the partial DOS of the central atoms only.

In summary, the presented simulated indentation experiments showed that fixed single MoS_2 layers show a robust behavior upon deflection for both mechanical and electronic properties. This robustness is a good condition for potential applications, such as flexible electronic devices. However, in the extreme cases, e.g. at the rupture point, this statement will not be valid.

ACKNOWLEDGMENTS

The authors acknowledge fruitful discussions with Andras Kis.

REFERENCES

1. R. Fivaz and E. Mooser, Phys. Rev. **163** (1967), 743.
2. P. Joensen, R. F. Frindt, and S. R. Morrison, Mater. Res. Bull. **21** (1986), 457.
3. K. S. Novoselov, D. Jiang, F. Schedin, T. J. Booth, V. V. Khotkevich, S. V. Morozov, A. K. Geim, Proc. Nat. Acad. Sci. U.S.A. **102** (2005), 10451.
4. T. Li and G. Galli, J. Phys. Chem. C **111** (2007), 16192.
5. B. Radisavljevic, A. Radenovic, J. Brivio, V. Giacometti, and A. Kis, Nature Nanotechnol. **6** (2011), 147.
6. I. Kaplan-Ashiri, S.R. Cohen, K. Gartsman, R. Rosentsveig, G. Seifert, and R. Tenne, J. Mater. Res. **19** (2004), 454.
7. S. Bertolazzi, J. Brivio, and A. Kis, ACS Nano **5** (2011), 9703.
8. D. Porezag, T. Frauenheim, T. Köhler, G. Seifert, and R. Kaschner, Phys. Rev. B **51** (1995), 12947.
9. G. Seifert, D. Porezag, and T. Frauenheim, Int. J. Quantum Chem. **58** (1996), 185.
10. G. Seifert and J.-O. Joswig, Wiley Interdisc. Rev.: Comput. Mol. Sci. **2** (2012), 456.
11. A. M. Köster, G. Geudtner, A. Gourşot, T. Heine, A. Vela, D. Salahub, and S. Patchkovskii, deMon; NRC: Ottawa, Canada, 2004.
12. A. Föppl and L. Föppl, *Drang und Zwang*, München (1920), 1st Ed., p. 175.
13. M. Neek-Amal and F. M. Peeters, Phys. Rev. B **81** (2010), 235421.
14. J. L. Feldman, J. Phys. Chem. Solids **37** (1976), 1141.

Mater. Res. Soc. Symp. Proc. Vol. 1549 © 2013 Materials Research Society
DOI: 10.1557/opl.2013.812

Graphene and The Advent of Other Layered-2D Materials for Nanoelectronics, Photonics and Related Applications

Anupama B. Kaul *(invited paper)*

Division of Electrical, Communications and Cyber Systems, Engineering Directorate,
National Science Foundation, Arlington VA 22203
Email: akaul@nsf.gov

ABSTRACT

Carbon-based nanostructures have been the center of intense research and development for more than two decades now. Of these materials, graphene, a two-dimensional (2D) layered material system, has had a significant impact on science and technology in recent years after it was experimentally isolated in single layers in 2004. The recent emergence of other classes of 2D layered systems beyond graphene has added yet more exciting and new dimensions for research and exploration given their diverse and rich spectrum of properties. For example, h-BN a layered material closest in structure to graphene, is an insulator, while NbSe, a transition metal dichalcogenide is metallic and monolayers of other transition metal di-chalcogenides such as MoS_2 are direct band-gap semiconductors. The rich variety of properties that 2D layered material systems offer can potentially be engineered on-demand, and creates exciting prospects for their device and technological applications ranging from electronics, sensing, photonics, energy harvesting and flexible electronics in the near future.

INTRODUCTION

It is well known that carbon-based nanomaterials such as graphene and carbon nanotubes exhibit remarkable mechanical, electrical, thermal and optical properties which has stirred a great deal of excitement for considering them for a wide variety of applications ranging from nanoscale transistors,[1,2,3] interconnects,[4] ultra-capacitors,[5] biosensors,[6] stretchable electronics,[7] thermo-electrics,[8] photo-voltaics,[9,10] optical applications and plasmonics,[11,12] as well as nano-electro-mechanical-systems (NEMS)[13,14,15,16,17] given their remarkable mechanical properties.[18,19,20] The investigation of graphene as a model two-dimensional (2D) system has impacted a diverse array of fields spanning physics, chemistry, materials science, and engineering.

While great strides have been made recently for applications of graphene that have stemmed from its unique properties, the absence of an intrinsic band-gap in pristine graphene poses concerns for its attractiveness in electronics applications, specifically digital electronics, where high ON/OFF ratios are desired. Although a band-gap in graphene is induced through quantum confinement by creating graphene nanoribbons (GNRs),[21] chemical functionalization,[22] and through the application of an electric field in bilayer graphene,[23] the band gaps nonetheless are small (few hundred meV). Recently, layered 2D crystals of other materials similar to graphene have been realized which include the transition metal di-chalcogenides, transition metal oxides and other 2D compounds such as insulating hexagonal-BN, Bi_2Te_3 and Bi_2Se_3. In this paper, we will start with an overview of graphene which has enabled a transformational impact on both science and technology over the past 8 or 9 years since it was first isolated experimentally in 2004. We will then introduce other layered 2D layered nanomaterials, such as MoS_2, which are beginning to play an important role for enabling innovative device applications in electronics, photonics, sensing, energy harnessing, flexible electronics and other related applications.

GRAPHENE: STRUCTURE & PROPERTIES

For more than two decades, carbon-based nanostructures have been intensely explored by the research community given their remarkable electrical, mechanical, optical and thermal properties. Such materials include the exploration of carbon nanotubes (CNTs), carbon nanofibers (CNFs), Y-junctions, graphene and graphene nanoribbons (GNRs), which either enable new device functionality or enhance the performance of established architectures in electronics.

The origin of the extraordinary materials properties in carbon-based nanomaterials arises from the structural arrangement of carbon atoms in the crystalline lattice. While carbon possesses a diverse spectrum of allotropes, ranging from 3D diamond and graphite (Fig. 1a), to quasi-one-dimensional (1D) single-walled-carbon-nanotubes (SWCNTs), multi-walled- (MW)-CNTs, carbon nanofibers, to zero-dimensional (0D) bucky balls, the 2D form of carbon, or for that matter the 2D form of any material, has remained elusive to experimentalists until recently. Single sheets of graphene were a theoretical curiosity until 2004 when Novoselov and Gheim succeeded in isolating a single sheet of carbon atoms from 3D graphite through mechanical exfoliation,[24] and their seminal experiments for the ensuing 6 years earned them the 2010 Nobel Prize in Physics.

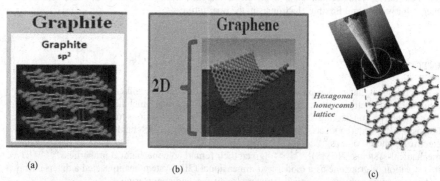

Figure 1. Crystal structures of (a) 3D graphite. A single layer of carbon atoms in graphite yields (b) 2D graphene. (c) The graphene lattice is a hexagonal honeycomb arrangement of carbon atoms. The lead pencil, which has been used for centuries, is composed of 3D graphite.

A schematic of this single sheet of carbon atoms is shown in Fig. 1b where the atoms are arranged in a hexagonal honeycomb crystal lattice. As Figure 1c illustrates, the hexagonal honeycomb lattice of 2D graphene represents the thinnest material we know of to date and yet, this membrane-like material is 5X stronger than steel owing to the strong in-plane sigma bonds, and yet it is far lighter than steel. Graphene is all surface and no bulk and exhibits a high mobility > 100,000 cm^2/V-s at room temperature and a high thermal conductivity of 5 x 10^3 W/m-K. Graphene has an ultra-high current carrying capability ~ 1 x 10^9 A/cm^2. Due to its flexibility, strength, high conductivity, transparency and low cost, graphene has been proposed as a replacement for indium tin oxide for solar cells,[25] and organic light emitting diodes, as well as in touch screens.[26] The large surface area to volume ratio of graphene suggests that it also has promise in ultra-capacitor applications.[27]

While graphene has been shown to exhibit exceptional promise for a wide range of applications, its lack of a band-gap as illustrated by the E-k diagram in Fig. 2a, poses serious concerns for its attractiveness in some applications, particularly digital electronics where high ON/OFF ratios are desired. Although one approach for inducing a band-gap in graphene is through quantum confinement by creating GNRs as shown in Fig. 2b with widths < 10 nm, the band gaps nonetheless are small (< few hundred meV), and it is challenging to maintain pristine edge chirality due to defects that are induced during nanofabrication of the ribbons. The methods used to induce a band-gap in graphene increase complexity and reduce the mobilities that pristine graphene offers.

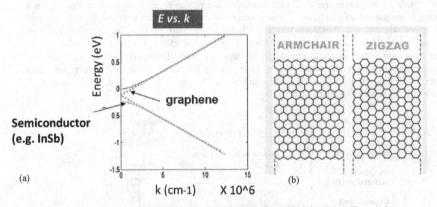

Figure 2. (a) E-k diagram for pristine graphene shows that it does not possess an intrinsic band gap unlike typical semiconductors (e.g. InSb shown here) which exhibit a spectral gap. (b) GNR exhibiting the arm-chair and zigzag chirality. It is difficult to maintain pristine edges in forming GNRs through nanofabrication techniques.

OTHER 2D LAYERED MATERIALS BEYOND GRAPHENE

Recently, layered 2D crystals of other materials similar to graphene have been realized which include insulating hexagonal-BN (band gap ~5.5 eV) and transition metal di-chalcogenides which display properties ranging from metallic NbS_2 to semiconducting MoS_2. The transition

13

metal di-chalcogenides consist of hexagonal layers of metal M atoms sandwiched between two layers of chalcogen atoms X with stoichiometry MX_2 as shown in Fig. 3 for the case of MoS_2 (M = Mo, X = S). As with transition metal di-chalcogenides in general, the interatomic binding in MoS_2 is strong arising from the covalent in-plane bonding but the subsequent layers interact through the weaker van der Waals interlayer forces.

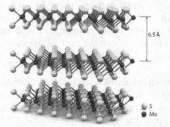

Figure 3. Crystalline structure of layered MoS_2 where the Mo atom is sandwiched between the S atom. The interlayer bonding occurs via the weak van der Waals interaction. Unlike graphene, monolayers of MoS_2 exhibit a large band-gap ~ 1.8eV and it is also a direct-band gap semiconductor.

Depending on the combination of the transition metal atom and the chalcogen (S, Se or Te), a wide variety of transition metal di-chalcogenides are possible, as illustrated in Fig. 4, each offering a unique set of properties. The coordination and oxidation state of the metal atoms determines whether the transition metal di-chalcogenide will be metallic, semi metallic or semiconducting. Superconductivity and charge density wave effects have also been observed in some transition metal di-chalcogenides. Besides the transition metal dichalcogenides, the chalcogenides of group III (GaSe, GaTe, InSe), group IV (GeS, GeSe, SnS, SnSe, etc.) and group V (Bi_2Se_3, Bi_2Te_3) also show a graphite like layered structure and offer promise in electronics, photonics and energy harvesting.

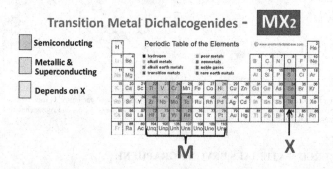

Figure 4. The transition metal di-chalcogenides are an example of 2D layered materials. Depending on the combination of the transition metal M, and the chalcogen atom X (S, Se, Te), a wide range of properties can arise.

Recently, it has been shown that bulk MoS_2 films are indirect band-gap semiconductors with a band gap of ~1.2 eV and a transformation takes place to a direct band gap semiconductor with a gap of ~1.8 eV for single-layers. Already, the device applications of such systems show promising characteristics; for example transistors derived from 2D monolayers of MoS_2 show ON/OFF ratios many orders of magnitude larger than the best graphene transistors at room temperature, with comparable mobilities.[28] The intrinsic band gap in these layered materials can be tuned depending on the choice of materials and implies possible applications in photonics, sensing and energy harvesting. The field of 2D materials beyond graphene is likely to grow at a rapid pace in the near future and while the many device applications that have emerged recently have been reported for mechanically exfoliated layers, progress in the nanomanufacturable synthesis of these materials will prove to be a key factor for propelling this field forward in the coming years.

CONCLUSIONS

An overview of graphene was presented along with a description of 2D layered materials including the transition metal di-chalcogenides. The transition metal di-chalcogenides represent a diverse source of 2D systems with exotic electronic and optical properties. The ability to engineer the materials properties in these 2D layered materials provides promising prospects for their use in a wide variety of applications ranging from electronics, photonics, sensing, energy harvesting, flexible electronics and related applications in the near future.

ACKNOWLEDGEMENTS

ABK wishes to acknowledge support for this through the NSF independent research and development (IR&D) plan.

Disclaimer: Any opinion, findings, and conclusions or recommendations expressed in this material are those of the author and do not necessarily reflect the views of the National Science Foundation.

REFERENCES

1   P. Avouris, Z. Chen, and V. Perebeinos, *Nat. Nanotechnol.* **2**, 605 (2007).
2   Q. Cao and J. Rogers, *Adv. Mater* **21**, 29 (2009).
3   A. Bachtold, P. Hadley, T. Nakanishi, and C. Dekker, *Science* **294**, 1317 (2001).
4   H. Li, C. Xu, N. Srivastava, and K. Banerjee, *IEEE Trans. Elect. Dev.* **56**, 1799 (2009).
5   H. M. Jeong, et al. Nano Lett. **11**, 2472 (2011).
6   F. Lu, L. Gu, M. J. Meziani, X. Wang, P. G. Luo, L. M. Veca, L. Cao, and Y. P. Sun, *Adv. Mater.* **21**, 139 (2009).
7   K. S. Kim, Y. Zhao, et al. *Nature* **457**, 706 (2009).
8   P. Wei, W. Bao, Y. Pu, C. N. Lau, and J. Shi, *Phys. Rev. Lett.* **102**, 166808 (2009).
9   X. Miao, S. Tongay, M. K. Petterson, K. Berke, A. G. Rinzler, B. R. Appleton, and A. F. Hebard, *Nano Lett.* **12**, 2745 (2012).
10   X. Dang, H. Yi, M. Ham, J. Qi, D. Yun, R. Ladewski, M. S. Strano, P. T. Hammond, and A. M. Belcher, *Nature Nano.* **6**, 377 (2011).
11   Y. Homma, S. Chiashi, and Y. Kobayashi, *Reports on Progress in Physics* **72**, 066502 (2009).
12   A. Vakil and N. Engheta, *Science* **332**, 1291 (2011).

13 J. E. Jang, S. N. Cha, Y. J. Choi, D. J. Kang, T. P. Butler, D. G. Hasko, J. E. Jung, J. M. Kim, and G. A. J. Amaratunga, *Nat. Nanotech.* **3**, 26 (2008).

14 O. Y. Loh and H. D. Espinosa, *Nature Nano.* **7**, 283 (2012).

15 A. B. Kaul, E. W. Wong, L. Epp, and B. D. Hunt, *Nano Lett.* **6**, 942 (2006).

16 A. B. Kaul, A. Khan, L. Bagge, K. G. Megerian, H. G. LeDuc, and L. Epp, *Appl. Phys. Lett.* **95**, 093103, (2009).

17 A. B. Kaul, K. Megerian, P. von Allmen, R. L. Baron, *Nanotechnology* **20**, 075303 (2009)

18 M. F. Yu, O. Lourie, M. J. Dyer, K. Moloni, T. F. Kelly, and R. S. Ruoff, *Science* **287**, 637 (2000).

19 C. Lee, X. Wei, J. W. Kysar, and J. Hone, *Science* **321**, 385 (2008).

20 A. B. Kaul, K. G. Megerian, A. Jennings, and J. R. Greer, *Nanotechnology* **21**, 315501 (2010).

21 X. Li, X. Wang, L. Zhang, S. Lee, and H. Dai, *Science* **319**, 1229–1232 (2008)

22 Balog, R. *et al. Nature Mater.* **9**, 315–319 (2010).

23 Zhang, Y. *et al. Nature* **459**, 820–823 (2009).

24 K. S. Novoselov, A. K. Geim, S. V. Morozov, D. Jiang, Y. Zhang, S. V. Dubonos, I. V. Grigorieva, and A. A. Firsov, *Science* **306**, 666 (2004).

25 X. Wang, Z. Zhi, and K. Mullen, *Nano Lett.* **8**, 323 (2009).

26 P. Matyba, H. Yamaguchi, G. Eda, M. Chhowalla, L. Edman and N. D. Robinson, *ACS Nano* **4**, 637 (2010).

27 M. D. Stoller, S. Park, Y. Zhu, J. An and R. S. Ruoff, *Nano Lett.* **8**, 3498 (2008).

28 B. Radisavljevic, A. Radenovic, J. Brivio, V. Giacometti, and A. Kis, *Nature Nano* **6**, 147 (2011).

Graphene

Mater. Res. Soc. Symp. Proc. Vol. 1549 © 2013 Materials Research Society
DOI: 10.1557/opl.2013.964

Chemical stability of epoxy functionalizations of graphene: A density functional theory study

Si Zhou[1,2] and Angelo Bongiorno[1]

[1]School of Chemistry & Biochemistry, Georgia Institute of Technology, Atlanta, GA 30332-0400, U.S.A.

[1]School of Physics, Georgia Institute of Technology, Atlanta, GA 30332-0430, U.S.A.

ABSTRACT

Density functional theory and statistical calculations are combined to address the chemical stability and structure of epoxy functionalizations of single-layer graphene. Our computations show that at oxidation levels of O:C<0.5, the Gibbs free energy of formation per epoxide amounts to about 0.6 eV, and the structure of the epoxy functionalizations presents local order and long-range disorder. The positive energy value indicates that in air at p=1 bar and room temperature, epoxy functionalizations of graphene are unstable and prone to spontaneous reduction. Our calculations show also that formation and release of O_2 is a slow process whose kinetics is controlled by large energy barriers, the formation of very stable intermediate species, and unlikely electronic transitions.

INTRODUCTION

Known for over 150 years [1], oxidized graphite, which can be exfoliated into two-dimensional graphene oxide, is the most studied chemically modified form of graphene [2]. This form of graphene oxide is obtained by using harsh treatments [3], leading to a carbon material with a complex chemistry and structure, and whose properties are difficult to control, even by post-synthesis chemical [4] or thermal treatments [4]. More recent methods to produce high-quality graphene oxide films are based on the use of epitaxial graphene on SiC(0001) and well-controlled physical [5] and chemical [6] oxidation methods. These new forms of ultra-thin oxidized epitaxial graphene have been shown to consist of well-structured carbon films, presenting a homogeneous distribution of predominantly epoxide and hydroxyl groups [5, 6]. Thanks to their reduced complexity, these new forms of graphene oxide films open the way to fundamental studies and novel applications of chemically functionalized graphene. In this work, we use density functional theory (DFT) calculations to address the chemical stability and structure of epoxy functionalizations of graphene, recently obtained by using atomic oxygen in ultrahigh vacuum [5].

Ordered and fully oxidized structures of graphene oxide have been investigated in great detail from DFT to date [7–9], while only a few computational efforts have focused on the disordered and partially oxidized phases of this material [6, 10]. In this work, we combine DFT and statistical models to investigate the chemical stability and structure of oxidized epitaxial graphene presenting realistic O:C ratios and irregular spatial distributions of oxygen functional groups. Here, we disregard the multilayer geometry of the film and the role of the substrate, and we focus only on epoxy functionalizations of single-layer graphene.

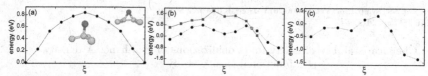

Figure 1. Symbols show energy profiles obtained from NEB DFT calculations of (a) an epoxide migrating between two nearest neighbor C-C bonds, and two epoxide groups reacting to form (b) a free O_2 molecule and (c) a carbonyl-pair species. Insets in (a) show stable (left) and transition-state (center) configurations of a migrating epoxide group. Similar configurations for reactions (b) and (c) are shown in Fig. 2. In (b), reaction energy paths obtained from spin-polarized DFT calculations with total spin moment equal to zero (singlet state) and one (triplet state) are shown by using black discs and blue squares connected by dotted segments, respectively. The circle shows the interception between the two energy profiles.

METHODS

In this work, we perform total energy and nudged elastic band (NEB) DFT calculations by using the QUANTUM-Espresso package [11]. We use norm-conserving pseudopotentials [12] for both the O and C species, a plane-wave energy cutoff of 65 Ry, and the Perdew-Burke-Ernzerhof [13] parametrization of the exchange and correlation energy functional. To model single-layer graphene, we use a supercell with 5×6 unit cells of graphene in the xy-plane, separated by vacuum region in the z direction. This tetragonal supercell includes 60 C atoms and has dimensions 12.36 Å $\times$ 12.84 Å $\times$ 12 Å.

To calculate the energy of formation, ΔE^1, of a single epoxide on graphene, we use the following equation:

$$\Delta E^1 = E_{DFT}(G + epoxide) - E_{DFT}(G) - E_{DFT}(O_2)/2, \qquad (1)$$

where the three energy terms on the right side correspond to the total DFT energy of, from left to right, defected graphene, pristine graphene, and O_2, respectively. Γ-point DFT calculations based on the aforementioned details give a ΔE^1 equal to 0.98 eV. DFT calculations based on denser k-point meshes in the Brillouin zone lead to energy values for ΔE^1 differing by only ± 0.05 eV. In the following, we report results obtained from Γ-point DFT calculations. We also remark that our DFT scheme gives dissociation energies and bond lengths for O_2 and CO of about 5.2 eV and 1.25 Å, and 10.8 eV and 1.13 Å, respectively. These data compare well with the experiment, and set the overall accuracy of our DFT results to about 3%.

RESULTS

Figure 1 shows the energy profiles obtained from NEB DFT calculations of an epoxide group migrating between two nearest neighbor C-C bonds (Fig. 1(a)) of graphene, two epoxide groups on the same side of a graphene layer reacting to form a free O_2 molecule (Fig. 1(b)), and two epoxide groups chemisorbed on opposite sides of a graphene layer reacting to form a carbonyl-pair species (Fig. 1(c)) [14]. These calculations show that the activation energy associated to epoxide migration on graphene is about 0.8 eV, and that similar reaction energy barriers are involved in the formation of both O_2 and the carbonyl-pair species (Fig. 1). Also, we find that gaseous O_2 in the triplet spin state is more stable than two isolated epoxides on graphene by about 1.8 eV (Fig. 2). When separated by only a few C-C bonds (Fig. 2), however, the two epoxide groups can find configurations exhibiting a slightly increased energetic stability, between 1.3 eV and 1.6 eV larger than the energy of free O_2. The carbonyl-pair species forms, with respect to epoxides, an equally very stable complex in graphene, lower by about 1.3 eV than the energy of two isolated epoxide groups, and larger by only 0.5 eV than the energy of free O_2. Further NEB DFT calculations show that the conversion of a carbonyl-pair species to a free O_2 involves a reaction energy barrier of about 2.8 eV, thereby making this process unlikely to occur at room temperature.

We use spin-polarized DFT calculations to locate the crossing point of the singlet and triplet potential energy surfaces associated with the reaction process shown in Fig. 1(b). Based on the Landau-Zener theory [15], we thus use the computed reaction energy profiles to estimate the probability, p_{s-t}, for singlet-to-triplet spin conversion at the transition state shown in Fig. 1(b). In particular, we use:

$$p_{s-t} = 2 \left[1 - \exp\left(-\frac{V^2}{hv|F_s - F_t|} \right) \right],
\tag{2}$$

where V is the spin-orbit matrix element between the triplet and singlet states of free O_2, v is the velocity of the O_2 center of mass at the transition state, and F_s and F_t are the forces at the transition state acting on a O_2 molecule in the singlet and triplet state, respectively. As in Ref. [16], we employ $V = 122$ cm^{-1} and v equal to root mean square velocity of a gaseous O_2 molecule at T=300K. F_s and F_t, on the other hand, are derived from the potential energy surfaces shown in Fig. 1(b) by taking the gradient with respect to the distance between the graphene layer and the O_2 center of mass. With the above values for V, v, and F_s and F_t, Eq. 2 gives a probability value of $p_{s-t} = 0.0042$, indicating a poor rate of spin conversion per single reaction process.

Figure 2 shows that the energetic interaction of two neighboring epoxide groups is either attractive or repulsive, depending on their mutual separation and whether the two species are chemisorbed on the same or opposite side of the carbon basal plane. To investigate aggregation phenomena, chemical stability, and structural properties of epoxy functionalizations of graphene, we use the results reported in Fig. 2 to devise a simplified statistical scheme. In particular, we resort to a lattice-model representation of functionalized graphene (Fig. 3), where epoxide groups on a 2D honey-comb lattice interact via the discrete pairwise energies shown in Fig. 2. Given a distribution of N epoxide groups on a periodic honey-comb lattice, we thus compute the Gibbs free energy of formation of the selected functionalization as:

$$\Delta G^N = \sum_{i>j} \epsilon_{ij} + N(\Delta E^1 - \Delta\mu),
\tag{3}$$

where ΔE^1 is defined in Eq. 1, ij and ϵ_{ij} refer to the epoxy-dimer complexes and relative energies shown in Fig. 2, and $\Delta\mu = -0.29$ eV [17] accounts for the entropic and enthalpic terms of the chemical potential of gaseous O_2 at p=1 bar and T=300 K. To explore the configurational space and determine optimal structures, we then use a standard Monte-Carlo approach. In this simulation scheme, epoxide groups are allowed to jump freely between nearest neighbors C-C bonds, and a Monte Carlo step consists of one of such "jumps". A single Monte Carlo step is accepted based on the Boltzmann factor weighting the relative statistical likelihood of initial and final configurations of the epoxide groups. Results obtained by using lattice models of epoxy functionalizations of graphene and Monte Carlo simulations are shown in Fig. 3.

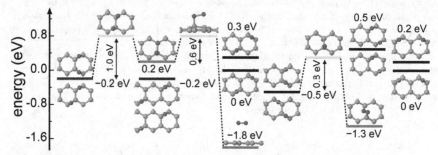

Figure 2. Line segments show the energy of two epoxide groups chemisorbed on a carbon basal plane, as illustrated in the ball and stick images above or below the segments. C and O atoms are shown in gray and red colors, respectively. Colored line segments indicate the transition and final states involved in the formation of a O_2 molecule (left) and a carbonyl-pair species (right). Energy values are referred to the energy of two isolated epoxide groups on graphene.

The sum of selected two-body energy terms in Eq. 3 corresponds to a convenient but also approximate way to calculate the energy of arbitrary epoxy functionalizations of graphene. To determine the accuracy of such an approximation, we consider a selected set of complexes formed by three neighboring epoxide groups and one crystalline phase of fully oxidized graphene, and we compare the results obtained by using Eq. 3 to those derived directly from DFT calculations (Fig. 3(a)). The comparison shows that the use of Eq. 3 introduces errors of about ±0.2 eV per epoxide with respect to the energies obtained directly from DFT. Nonetheless, Eq. 3 is capable of describing qualitative trends and, most importantly, the relative energy stability of the fully oxidized crystalline phase and small agglomerates of epoxide groups.

We use Eq. 3 and lattice model Monte Carlo simulations to investigate (at a qualitative level) the chemical stability and structure of epoxy functionalizations of single-layer graphene. To this end, we considered O:C ratios up to 0.5 and standard simulated annealing Monte Carlo methods to determine optimal structure and their Gibbs free energy of formation. In these simulations, epoxide groups are distributed at random on the graphene layer, and after a long equilibration at high temperature, the system is slowly quenched to room temperature and further equilibrated until the epoxy functionalization reaches an arrested state. An extended set of simulations leads to the results shown in Fig. 3(b). The simula-

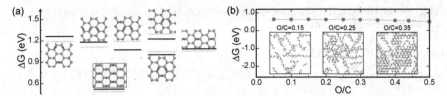

Figure 3. (a) Gibbs free energy of formation (per epoxide) of selected epoxy functionalizations of graphene obtained by using Eq. 3 (colored line segments) and straightforward DFT computations (black line segments). The illustrations reported above or below the line segments show the selected complexes formed by three neighboring epoxide groups, as well as the most stable crystalline phase of fully epoxy-oxidized graphene layer (colored box). (b) Gibbs free energy of formation per epoxide vs. O:C ratio of functionalized graphene presenting irregular distributions of epoxide species chemisorbed on both sides of the carbon plane. Symbols show results obtained by using Eq. 3 and simulated annealing Monte Carlo simulations. The magenta colored disc shows the energy of the crystalline phase shown in (a). Insets show selected epoxy functionalizations of graphene obtained from these simulations. Red and yellow discs show epoxide groups chemisorbed on the two opposite sides of the carbon layer. The honey-comb lattice of graphene is not shown.

tions show that the Gibbs free energy of formation of epoxy functionalizations of single-layer graphene varies little with the O:C ratio, slowly decreasing from an energy value of 0.64 eV at O:C=0.05 to 0.55 eV at O:C=0.5, at which the graphene layer is fully oxidized and the epoxide species are distributed regularly as shown in Fig. 3(a). In the case of partially oxidation, that is O:C<0.5, our simulations show that the spatial distribution of the functional groups presents no long-range order and only some degree of short-range order. In spite of the partial coverage and lack of long-range order, epoxy functionalizations of graphene presenting local order achieve a chemical stability very close to that of the fully oxidized and ordered phase (Fig. 3).

CONCLUSIONS

DFT calculations show that the migration of epoxide species chemisorbed on graphene involves energy barriers of about 0.8 eV. Nearest-neighbor epoxides can react to form O_2 molecules by overcoming energy barriers of 0.6-0.8 eV, and up to 2.8 eV when the reaction occurs via formation of a carbonyl-pair species. At room temperature, O_2 release is thus a process kinetically unfavored, further slowed down by the poor rate of spin exchange at the crossing of the singlet and triplet energy surfaces. The Gibbs free energy of formation of a single epoxide on graphene amounts to about 0.98 eV. At oxidation levels such that O:C<0.5, the Gibbs free energy of formation per epoxide decreases to about 0.6 eV, and the structure of the epoxy functionalizations presents local order and long-range disorder. In spite of local ordering, the formation of energetically favored motifs, and thus an enhanced energetic stability, epoxy functionalizations of graphene retain a positive value of $\Delta_f G$, indicating that in air at p=1 bar and room temperature this material is unstable and prone to spontaneous reduction. Formation and release of O_2 is, however, a process whose kinetics is controlled

23

by large energy barriers, the formation of very stable intermediate species (carbonyl-pairs), and unlikely electronic transitions.

ACKNOWLEDGEMENTS

The authors acknowledge support from the NSF grants CMMI-1100290, DMR-0820382, and CHE-0946869.

REFERENCES

[1] B. Brodie, Ann. Chim. Phys. **45**, 351 (1855).

[2] S. Park and R. S. Ruoff, Nat. Nanotechnol. **4**, 217 (2009).

[3] W. S. Hummers and R. E. Offeman, J. Am. Chem. Soc. **80**, 1339 (1958).

[4] S. Stankovich, D. A. Dikin, R. D. Piner, K. A. Kohlhaas, A. Kleinhammes, Y. Jia, Y. Wu, S. T. Nguyen, and R. S. Ruoff, Carbon **45**, 1558 (2007).

[5] M. Z. Hossain, J. E. Johns, K. H. Bevan, H. J. Karmel, Y. T. Liang, S. Yoshimoto, K. Mukai, T. Koitaya, J. Yoshinobu, M. Kawai, A. M. Lear, L. L. Kesmodel, S. L. Tait, and M. C. Hersam, Nature Chem. **4**, 305 (2012).

[6] S. Kim, S. Zhou, Y. Hu, M. Acik, Y. Chabal, C. Berger, W. de Heer, A. Bongiorno, and E. Riedo, Nature Mat. **11**, 544 (2012).

[7] D. Boukhvalov and M. Katsnelson, J. Am. Chem. Soc. **130**, 10697 (2008).

[8] J.-A. Yan and M. Y. Chou, Phys. Rev. B **82**, 125403 (2010).

[9] P. Johari and V. Shenoy, ACS Nano **5**, 7640 (2011).

[10] J. T. Paci, T. Belytschko, and G. C. Schatz, J. Phys. Chem. C **111**, 18099 (2007).

[11] P. Giannozzi *et al.*, J. Phys.: Condens. Matter **21**, 395502 (2009).

[12] N. Troullier and J. L. Martins, Phys. Rev. B **43**, 1993 (1991).

[13] J. P. Perdew, K. Burke, and M. Ernzerhof, Phys. Rev. Lett. **77**, 3865 (1996).

[14] A. Bagri, C. Mattevi, M. Acik, Y. J. Chabal, M. Chhowalla, and V. B. Shenoy, Nature Chem. **2**, 581 (2010).

[15] C. Zener, Proc. R. Soc. Lond. A **137**, 696 (1932).

[16] W. Orellana, A. J. R. da Silva, and A. Fazzio, Phys. Rev. Lett. **90**, 016103 (2003).

[17] D. R. Stull and H. Prophet, *JANAF Thermochemical Tables* (U.S. National Bureau of Standards, Washington, D.C., ADDRESS, 1971).

Mater. Res. Soc. Symp. Proc. Vol. 1549 © 2013 Materials Research Society
DOI: 10.1557/opl.2013.1029

Graphene Oxide-Supported Two-Dimensional Microporous Polystyrene

Yi Ouyang, Dingcai Wu and Ruowen Fu[*]
Materials Science Institute, PCFM Lab and DSAPM Lab, School of Chemistry and Chemical
Engineering, Sun Yat-sen University, Guangzhou 510275, P. R. China

ABSTRACT

In this paper, a microporous-containing graphene oxide/polystyrene (M-GO/PS) was
designed and prepared by surface-initiated atom transfer radical polymerization (SI-ATRP) of PS
from GO surface and then crossrlinking by carbon tetrachloride. The structures of the molecular
brush of PS and the related crosslinking M-GO/PS were determined by FTIR, TG, SEM and
nitrogen adsorption-desorption analysis. The experimental results showed that PS molecular
brush were successfully grown on to the surface of GO. After crosslinking, the PS component
was crosslinked into many round nanoparticles with a diameter of 20-30 nm, and therefore the
specific surface area of GO/PS obviously increased. This kind of porous M-GO/PS composite
was promising for the application in adsorption-desorption energy storage areas.

INTRODUCTION

Construction graphene, with a single layer plane structure, has amazing electrical and
mechanical property, such as high electroconductibility, flexibility, stiffness and etc.[1] This
unique structure has rendered graphene highly promising for various applications in energy
storage[2-4], electronic device[5,6]. To further explore new functions for graphene, the construction of
macroscopic architectures using graphene as the building block has been well performed. 0-
dimensional[7], two-dimensional and three-dimensional macrostructure have been constructed via
vacuum filtration[8], layer-by-layer (LBL) assembly[9] and chemical vapor deposition (CVD)[10].
Recently, great effort has been focused on porous graphene materials[11], Taking advantages of
both the graphene composition and the pore structures, porous graphene materials have great
application potentials in various fields. Three-dimensional porous graphene materials with high
surface area have been prepared by LBL assembly. However, previously reported researches are
limited in graphene foam. The design and preparation of graphene/ microporous structure is still
empty.

In this paper, a GO-*g*-PS plane molecular brush was designed and prepared by surface-
initiated atom transfer radical polymerization (SI-ATRP) of polystyrene (PS) from graphene
oxide (GO) surface at first. After hypercrosslinking of PS component, a microporous-containing
composite with GO/microporous PS (M-GO/PS) was obtained, as seen in Figure1. This newly-
obtained unique structure, i.e., micropores on plane surface, will render porous graphene material
highly promising application in the areas such as separation and adsorption/desorption.

GO   GO-g-PS   M-GO-PS
Figure 1. Illustration of preparation of graphene oxide supported-microporous polystyrene

EXPERIMENT

Preparation of GO-OH. GO-OH was prepared by modified Hummers method[12]. 500mg of dried GO was placed in a two-necked flask with condensation tube. Thionyl chloride was dropwise added into the system until no bubble can be seen from the surface of GO flake, and then the excessive thionyl chloride was moved by vacuum pump. 15 mL of glycol was added into the flask. Then the suspension was heated at 75 °C for 24 hours, filtrated, washed with ethanol and water for 3 times and finally dried at 50 °C under vacuum.

Preparation of GO-Br. The as-prepared GO-OH was dispersed in 20 mL of dried methylene dichloride. 1 mL of 2-Bromoisobutyryl bromide was dropwise added into the solution. Then, it was cooled to zero degree and kept for 3 hours.

Preparation of GO-g-PS. The GO-Br was dispersed in 20 mL of dried DMF. 0.27 mmol of CuBr and 0.27 mmol of PMDETA were added. After purging the reaction system with N_2 for 30 min, 28 mmol of styrene was added and the reaction temperature was raised to 80 °C. The polymerization was carried out for 24 hours. GO-g-PS was isolated by three times of successive centrifugation/redispersion using $CHCl_3$.

Crosslinking of GO-g-PS. 500mg of GO-g-PS was dispersed in 30 mL of CCl_4. 1.4g of $AlCl_3$ was then added. The mixture was heated at 75 °C for 24 hours, and then filtered, washed with acetone/HCl for 3 times, and finally dried at 50 °C.

Characterization. The grafting amount of PS from GO surface was calculated by using thermogravimetric analysis (TGA Q50) from 80 to 900 °C under nitrogen atmosphere with a heating rate of 10 °C/min. Gel Permeation Chromatograph (GPC, Waters Alliance GPC 2000) was used to determine molecular weight and its distribution of grafting PS chains which were cut down form GO by hydrolysis. The nanostructures of the samples were observed by a Hitachi S-3400 scanning electron microscope (SEM). N_2 adsorption measurement was carried out using a Micromeritics ASAP 2012 analyzer at 77 K. The BET surface area (S_{BET}) was determined by Brunauer-Emmett-Teller (BET) theory. The total pore volume (V_t) was estimated from the amount adsorbed at a relative pressure P/P_0 of ca. 0.99. The pore size distribution was analyzed by original density functional theory (DFT) combined with non-negative regularization and medium smoothing.

RESULTS AND DISCUSSION

Following the procedures described in the literature[13], the free carboxylic acid groups of GO were converted to acyl chloride groups , which upon treatment with ethylene glycol, yielded graphene sheets rich in hydroxyl groups (GO-OH) . After separated and freeze dried by lyophilisation, a feathery powder of GO-OH was obtained. Observing by SEM (as shown in Figure 2A), we can find the well exfoliated GO-OH sheets. The as-prepared GO-OH sheet was curl-up and had a smooth plane which consists of few layers.

In order to graft PS onto GO by ATRP, the ATRP initiator was induced at first, and the intermediate product (GO-Br) was also observed by TEM. It can also be seen that the GO sheets are still smooth (Figure 2 B) similar to that of GO-OH.

After grafting of PS from GO surface, the reaction mixture changed from yellow-brown to a black-gray color, indicating the conversion of GO-OH to GO-g-PS. The suspension status of GO-OH in the chloroform/water solution was shown in the inset of Figure 2A. The GO-OH was concentrated in aqueous phase, which clearly indicates the hydrophilic nature of GO-OH. In contrast, the as prepared GO-g-PS was concentrated in oil phase (inset in Figure 2C), demonstrating a hydrophobic nature, which indicates the success graft of PS by SI-ATRP. The grafting amount of PS was determined by TGA analysis. According to the result in Figure 3A, about 35% percentage of PS was grafted from GO surface.

Figure 2. SEM images of (A) GO-OH, (B) GO-Br, (C) GO-g-PS, (D), (E) and (F) M-GO/PS

The chain length of PS grown from the surface of GO, as well as its molecular weight distribution, was determined by GPC. Prior to this, the chains were dismounted from the surface via hydrolysis following Goncalves's work[14]. The GPC curve obtained is illustrated in Figure 3B, which clearly proves that the polymerization is very well controlled, with a MWD (or PD) of 1.27, indicating that the PS molecular brush was successfully grown on to the surface of GO.

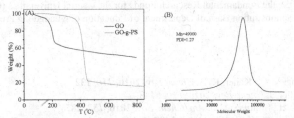

Figure 3. (A)TG curve of GO and GO-g-PS; (B) GPC curve of PS dismounted form GO-g-PS

The GO-g-PS was then treated with simple hypercrosslinking reaction based on Friedel-Crafts reaction of PS chains with the crosslinker of carbon tetrachloride (CCl_4). The M-GO/PS obtained had a weight gain of about 10% because of the introduction of -CO- crosslinking bridges which were translated from the crosslinker of CCl_4. The morphology of the M-GO/PS is showed in Figure 2D, E, F. After crosslinking, the PS component was crosslinked into many round nanoparticles with a diameter of 20-30 nm, and therefore the specific surface area of M-GO/PS obviously increased. According to the N_2 sorption isotherms determination (Figure 4), the M-GO/PS had a high nitrogen uptake at low relative pressure, indicating that it has a high microporosity, while the original GO sample has very low porosity. The general GO sheet has a very small surface area ($10.6\ m^2g^{-1}$). However, the specific surface area of the M-GO/PS is as high as $376\ m^2g^{-1}$ and its pore volume is as high as $0.5\ cm^3g^{-1}$ at $P/P_0=0.99$.

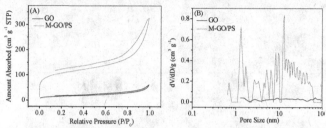

Figure 4. N_2 adsorption/ desorption isotherms (A) and pore size distribution curves (B) of GO and M-GO/PS

CONCLUSIONS

In summary, we have successfully prepared graphene-based porous materials. In contrast with ordinary porous graphene materials with foam structure, this entirely new material demonstrates a micropores-on-plane surface structure. Besides its plane nature, it also has high microporosity. Thus we believe that this two-dimensional microporous structure will find a spectrum of applications in areas such as separation and adsorption.

ACKNOWLEDGMENTS

We acknowledge financial support from the project of NSFC (51173213, 51172290, 50802116, 51232005), the Fundamental Research Funds for the Central Universities (09lgpy18), and the Project of Demonstration Base of Department of Education of Guangdong Province (cgzhzd0901).

REFERENCES

1. Allen, M. J.; Tung, V. C.; Kaner, R. B. *Chem. Rev.* **2010**, *110*, 132.
2. Nozik, A. J.; Miller, J. *Chem. Rev.* **2010**, *110*, 6443.
3. Du, A. J.; Zhu, Z. H.; Smith, S. C. *J. Am. Chem. Soc.* **2010**, *132*, 2876.
4. Geim, A. K.; Novoselov, K. S. *Nat. Mater.* **2007**, *6*, 183.
5. Wu, J. S.; Pisula, W.; Mullen, K. *Chem. Rev.* **2007**, *107*, 718.
6. Miller, R. D.; Chandross, E. A. *Chem. Rev.* **2010**, *110*, 1.
7. Zangmeister, C. D.; Ma, X. F.; Zachariah, M. R. *Chem. Mater.* **2012**, *24*, 2554.
8. Chen, H.; Muller, M. B.; Gilmore, K. J.; Wallace, G. G.; Li, D. *Adv. Mater.* **2008**, *20*, 3557.
9. Shen, J. F.; Hu, Y. Z.; Li, C.; Qin, C.; Shi, M.; Ye, M. X. *Langmuir* **2009**, *25*, 6122.

10. Chen, Z. P.; Ren, W. C.; Gao, L. B.; Liu, B. L.; Pei, S. F.; Cheng, H. M. *Nat. Mater.* **2011**, *10*, 424.
11. Lee, S. H.; Kim, H. W.; Hwang, J. O.; Lee, W. J.; Kwon, J.; Bielawski, C. W.; Ruoff, R. S.; Kim, S. O. *Angew. Chem. Int. Ed.* **2010**, *49*, 10084.
12. Hummers W S, O. R. E. *J. Am. Chem. Soc.* **1985**, *80*, 1339.
13. Spitalsky, Z.; Tasis, D.; Papagelis, K.; Galiotis, C. *Prog. Polym. Sci.* **2010**, *35*, 357.
14. Goncalves, G.; Marques, P. A. A. P.; Barros-Timmons, A.; Bdkin, I.; Singh, M. K.; Emami, N.; Gracio, J. *J. Mater. Chem.* **2010**, *20*, 9927.

Mater. Res. Soc. Symp. Proc. Vol. 1549 © 2013 Materials Research Society
DOI: 10.1557/opl.2013.1030

Discrete Gauge Fields for Graphene Membranes under Mechanical Strain

James V. Sloan,[1] Alejandro A. Pacheco Sanjuan,[2] Zhengfei Wang,[3] Cedric M. Horvath,[1] and Salvador Barraza-Lopez[1]
[1]Department of Physics. University of Arkansas. Fayetteville, AR 72701, USA,
[2]Departamento de Ingeniería Mecánica. Universidad del Norte. Barranquilla, Colombia,
[3]Department of Materials Science and Engineering. University of Utah. Salt Lake City, UT 84112, USA

ABSTRACT

Mechanical strain creates strong gauge fields in graphene, offering the possibility of controlling its electronic properties. We developed a gauge field theory on a honeycomb lattice valid beyond first-order continuum elasticity. Along the way, we resolve a recent controversy on the theory of strain engineering in graphene: there are no K-point dependent gauge fields.

INTRODUCTION

The interplay of electronic and mechanical properties of graphene membranes is a subject under intense experimental and theoretical investigation [1-7]. Mechanical strain induces gauge fields in graphene that affect the dynamics of charge carriers [3-6,8]. Graphene can sustain elastic deformations as large as 20% [9] resulting in pseudo-magnetic fields that are much larger than magnetic fields available in state-of-the-art experimental facilities [10]. These effects have been experimentally confirmed in strained graphene "nanobubbles" on a metal substrate [11]. In addition to the pseudo-magnetic vector potential As, strain induces a scalar deformation potential Es [8,12,13] that affects the electron dynamics in complex ways. Only in special cases Es=0 [4].

The theory of strain-engineered electronic effects on graphene is semiclassical. The strain-induced pseudo-magnetic field is given by the rotational of the strain-created vector potential: $Bs(r) = Rot[As(r)]$. The vector potential is incorporated into a spatially-varying pseudospin Hamiltonian Hps(q,r), where Hps(q) is the low-energy expansion of the Hamiltonian in reciprocal space in the absence of strain. The semiclassical approximation is justified when the strain extends over many unit cells and preserves sublattice symmetry [4]. It is also possible to determine the electronic properties directly from a tight-binding Hamiltonian H in real space, without resorting to the semiclassical approximation and without imposing an a priori lattice symmetry. So, while the semiclassical Hps(q,r) is defined in reciprocal space (thus assuming some reasonable preservation of crystalline order), the tight-binding Hamiltonian H in real space is more general.

We defer a discussion of specific details to forthcoming publications [14,15] and in the present occasion we focus on presenting our most relevant results, which include expressions for the strain-derived fields given directly in terms of changes between interatomic distances.

RELEVANT GEOMETRICAL PARAMETERS

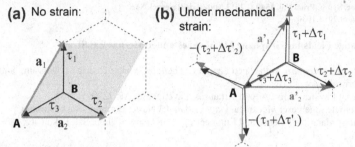

Figure 1. Lattice vectors and basis vectors change when graphene is under mechanical strain.

We show in Figure 1 the relevant information needed for our discussion. The choice of lattice vectors ($\mathbf{a}_1$ and $\mathbf{a}_2$) is the usual one [4,6], where the "zigzag" direction is parallel to the x-axis. (In reference 14 the zigzag direction lies along the y-axis instead.) There are three vectors joining nearest neighbors in the absence of strain, labeled by τ_1, τ_2, and τ_3. The unit cell in graphene (dashed area in Figure 1a) has two inequivalent atoms, labeled A and B. The distances among atoms change in the presence of mechanical strain (Figure 1b).

Let us first discuss the change on the lattice vectors: From Figure 1b we get:

$$\Delta\mathbf{a}_1 = \Delta\tau_1 - \Delta\tau_3, \text{ and } \Delta\mathbf{a}_2 = \Delta\tau_2 - \Delta\tau_3, \tag{1}$$

Now, we said that the theory of strain engineering is semiclassical. This means we must express a pseudospinor Hamiltonian at each unit cell. The pseudospinor Hamiltonians are objects defined in reciprocal (momentum) space, so the first part of our program should be to find out how reciprocal lattice vectors are renormalized upon strain.

There is a basic orthogonality relation among the lattice vectors and the reciprocal lattice vectors. Writing down the matrix A=($\mathbf{a}_1$,$\mathbf{a}_2$) in terms of the lattice vectors (written as column vectors), the relation between lattice and reciprocal lattice vectors B=($\mathbf{b}_1$,$\mathbf{b}_2$) is given by:

$$AB^T = 2\pi I, \tag{2}$$

where the superscript T indicates carrying a transpose operation, and I is the identity matrix. We find that, to first order on mechanical displacements, the reciprocal lattice vectors change as follows:

$$\Delta B = -2\pi(A^{-1}\Delta A A^{-1})^T, \tag{3}$$

where $\Delta A = (\Delta\mathbf{a}_1, \Delta\mathbf{a}_2)$.

Equation (3) is important for the following reason: The pseudospinor Hamiltonian is a low-energy expansion of the Hamiltonian in reciprocal space. The low-energy valence and conduction bands touch at six points in the Brillouin zone, usually labeled by an uppercase K.

These six high-symmetry points are related to $\mathbf{b}_1$ and $\mathbf{b}_2$ as follows:

$$K_1=(2\mathbf{b}_1+\mathbf{b}_2)/3, \; K_2=(\mathbf{b}_1-\mathbf{b}_2)/3, \; K_3=-(\mathbf{b}_1+2\mathbf{b}_2)/3, \; K_4=-K_1, \; K_5=-K_2, \text{ and } K_6=-K_3, \qquad (4)$$

Equations (3) and (4) tell us that if $\mathbf{b}_1$ and $\mathbf{b}_2$ changed due to strain, then the location of the K-points need to be changed as well. Extensive details can be found on publications [14,15].

RESULTS

Absence of K-point dependent velocity renormalization

In Reference 5, it is said that the theory needs to be modified to accommodate for K-point dependent gauge fields. In that reference the renormalization of reciprocal lattice vectors, Equation (3), is missing. When renormalization is included, no K-point dependent gauge fields appear. See Refs. 14, 15 for details. (See Refs. [16,17] as well.)

Expressions for gauge fields

These expressions can be used to obtain gauge fields directly from changes of atomic positions upon mechanical strain when the zigzag direction is oriented along the x-axis:

$$A_S=-(\beta\phi_0/3^{1/2}\pi a_0^3)[2\tau_3\,\Delta\tau_3 - \tau_1\,\Delta\tau_1 - \tau_2\,\Delta\tau_2 - 3^{1/2}i(\tau_2\,\Delta\tau_2-\tau_1\,\Delta\tau_1)] \qquad (5)$$

(see Ref. 14 for an expression holding when the zigzag direction is perpendicular to the x-axis). In the previous equation, a_0 is the lattice constant with no strain, $\beta\sim 2.3$, ϕ_0 is the magnetic flux quantum, and $i=(-1)^{1/2}$. Equation (5) tells us that it is possible to express gauge fields directly from atomic displacements. Equation (5) is significant because it can take atomistic information directly as its input. Equation (5) reduces to common expressions [4,6] when the changes in atomic positions due to strain are smaller than the lattice constant.

CONCLUSIONS

We showed in a simple way why no K-point dependent fields exist on a first-order theory of strain engineering in graphene, and provided expressions for the vector potential given solely in terms of atomic displacements caused by mechanical strain.

ACKNOWLEDGEMENTS

We thank M. Vanevic, and M.A. Kuroda for discussions, and computer support from HPC at Arkansas (RazorII), and XSEDE (TG-PHY090002, Blacklight at PSC, and Stampede at TACC).

REFERENCES

1. Meyer, J. C.; Geim, A. K.; Katsnelson, M. I.; Novoselov, K. S.; Booth, T. J.; Roth, S. *Nature* **446**, 60 (2007).
2. Bunch, J. S.; van der Zande, A. M.; Verbridge, S. S.; Frank, I. W.; Tanenbaum, D. M.; Parpia, J. M.; Craighead, H. G.; McEuen, P. L. *Science* **315**, 490 (2007).
3. Pereira, V. M.; Castro-Neto, A. H. *Phys. Rev. Lett.* **103**, 046801 (2009).
4. Guinea, F.; Katsnelson, M. I.; Geim, A. K. *Nature Physics* **6**, 30 (2010).
5. Kitt, A. L.; Pereira, V. M.; Swan, A. K.; Goldberg, B. B. *Phys. Rev. B* **85**, 115432 (2012).
6. Vozmediano, M. A. H.; Guinea, F. *Phys. Rep.* **496**, 109 (2010).
7. de Juan, F.; Sturla, M.; Vozmediano, M. *Phys. Rev. Lett.* **108**, 227205 (2012).
8. Suzuura, H.; Ando, T. *Phys. Rev. B* **65**, 235412 (2002).
9. Kim, K. S.; Zhao, Y.; Jang, H.; Lee, S. Y.; Kim, J. M.; Kim, K. S.; Ahn, J.-H.; Kim, P.; Choi, J.-Y.; Hong, B. H. *Nature* **457**, 706 (2009).
10. As of March of 2012, the highest reported magnetic field is of the order of 100 Tesla. See: http://www.magnet.fsu.edu/mediacenter/news/pressreleases/2012/100tshot.html
11. Levy, N.; Burke, S. A.; Meaker, K. L.; Panlasigui, M.; Zettl, A.; Guinea, F.; Castro-Neto, A. H.; Crommie, M. F. *Science* **329**, 544 (2010).
12. de Juan, F.; Cortijo, A.; Vozmediano, M. A. H. *Phys. Rev. B* **76**, 165409 (2007).
13. Choi, S.-M.; Jhi, S.-H.; Son, Y.-W. *Phys. Rev. B* **81**, 081407 (2010).
14. J. V. Sloan, A.A. Pacheco Sanjuan, Z. Wang, C.M. Horvath, and S. Barraza-Lopez, *Phys. Rev. B* **87**, 155436 (2013).
15. SBL, A. A. Pacheco-Sanjuan, Z. Wang, and M. Vanevic. *Solid State Comm.* **166**, 70 (2013).
16. F. de Juan, J. L. Mañes, M. A. H. Vozmediano, *Phys.Rev. B* **87**, 165131 (2013).
17. A. Kitt, V. M. Pereira, A. K. Swan, B. B. Goldberg, *Phys. Rev. B* **87**, 159909(E) (2013).

Mater. Res. Soc. Symp. Proc. Vol. 1549 © 2013 Materials Research Society
DOI: 10.1557/opl.2013.946

Electron Irradiation of Graphene Field Effect Transistor Devices

Sung Oh Woo[1] and Winfried Teizer[1, 2]
[1]Department of Physics and Astronomy, Texas A&M University, College Station, TX 77843-4242, U.S.A.
[2]WPI-Advanced Institute for Materials Research, Tohoku University, Sendai, Japan.

ABSTRACT

We report the effects of electron irradiation on graphene Field Effect Transistor (FET) devices. We irradiated the graphene devices with 30keV electrons and measured the electrical transport properties in high vacuum in-situ. Upon electron irradiation, a Raman 'D' band appears. In addition, we observed that the doping behavior of the graphene devices changed from P to N type as a result of the irradiation. We also observed a shift of the Dirac point while the graphene FET device stays in vacuum and after it interacted with environmental molecules under ambient conditions.

INTRODUCTION

Graphene, a single layer of carbon atoms, has attracted a great amount of attention not only due to scientific interest but also due to potential applications such as high speed electronic devices [1], molecular sensors [2], and molecule storage media [3]. In particular, graphene is considered a potential material for the post silicon era due to its extraordinary electronic properties. Because Scanning Electron Microscopy (SEM) and Electron Beam Lithography (EBL) are important tools in nanofabrication and characterization, Graphene Field Effect Transistor (GFET) devices might be routinely exposed to energetic electron beams. Therefore, it is important to study the effect of electron irradiation on graphene. We present the effects of electron irradiation on graphene FET devices, focusing specifically on its electronic transport properties.

EXPERIMENT

Graphene FET devices for electronic transport measurements were fabricated by the mechanical exfoliation technique. Graphene flakes were cleaved from 'YGA' grade highly oriented pyrolytic graphite and transferred onto substrates. Consisting of 285nm-think oxide layers grown on silicon wafers, these were selected in order to maximize contrast of the graphene flakes [4]. Single layer graphene flakes were then selected by optical microscopy and confirmed by Raman spectroscopy [5]. The electrodes for electronic transport measurements were patterned by EBL, and Cr(3nm)/Au(50nm) electrodes were deposited by thermal evaporation. The GFET was irradiated in high vacuum by an electron beam, with energy 30keV and current 10pA. For the AC transport measurement, a standard Lock in amplifier technique was employed. After electron irradiation, the transport measurement was conducted in-situ in the EBL chamber (JSM-6460, $2*10^{-5}$ Torr) without breaking vacuum. After the measurement was completed, Raman spectroscopy data (633nm excitation laser) was acquired to investigate the presence of 'defect' peaks in ambient conditions.

DISCUSSION

The characteristics of a typical GFET device prior to electron irradiation are shown in Figure 1. Figure 1(a) displays the optical image of the GFET device, the green dotted square being the electron irradiation area (size: 10um×20um). The Raman spectrum, in Figure 1(b), confirms that the pristine graphene is defect free because no Raman 'D' band appears. The Full Width Half Maximum (FWHM) of the Lorentzian fitting from the Raman 2D peak in the inset is ~27cm^{-1}, which confirms the graphene as a single layer of carbon atoms [5]. The unusually high background in the Raman shift region below 1700 cm^{-1} is related to the 2nd order phonon of silicon [6].

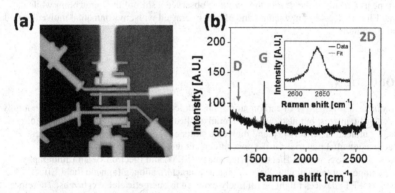

Figure 1. (a) Optical image of a GFET device. The green dotted square indicates the electron irradiation area (size: 10um×20um). (b) Raman data of the graphene device before irradiation. Single layer graphene is confirmed by a Lorentzian fit of 2D peak, with FWHM ~27cm^{-1}. The device shows no 'D' band around 1260cm^{-1}, which would be a signature of defects.

Resistivity data prior to the irradiation is displayed in Figure 2(a), which shows initial P-type doping behavior in ambient conditions. As the gas phase around the device is evacuated, the transport curve slightly shifts to lower gate voltages (V_g). Because adatoms or molecules physically attached to the graphene are removed from the surface, the Dirac point shifts slightly and the maximum resistivity is reduced. Annealing the graphene in vacuum shows similar results. Previous research reported that the maximum resistivity point shifts to lower V_g value as the device is annealed in vacuum [7], which is related to the removal of environmental surface molecules. Furthermore, the P-type doping behavior changes to N-type when annealing, which is in agreement with our observations. In this sense, both the vacuum and the annealing processes force some physically adsorbed air molecules or adatoms to desorb from the graphene surface. As a result, it is typically seen that adatoms contribute to the P-type doping behavior of graphene. After the electron irradiation of the graphene, the Raman 'D' band, which is measured after transport measurements, appeared, as shown in Figure 2 (b). This is an indication of defect creation. The Raman 'defect' signature appears when the symmetry of the $A1_g$ mode, a lateral vibration of carbon atoms in a honeycomb lattice, is broken. It occurs as a result of the displacement of carbon atoms from their lattice points or adsorption of molecules or atoms on graphene. In our experiment, it seems energetically impossible for electrons with typical energies around 30keV to eject carbon atoms from their original position because the threshold energy of such defects is more than 80keV [8]. Previous work reported that a Raman 'D' band appears due to irradiation of electrons with energy ranging from 5keV to 30keV [9-10]. Other work found that the 'D' band can emerge as a result of adsorption of single atoms [11] or fragments of environmental molecules such as H_2O [12] on graphene. Because energetically this is the sole remaining scenario, we propose that the Raman 'D' band originated from adsorption of fragmented molecules or atoms on the GFET device due to interaction between energetic electrons and carbon atoms. After an extended measurement of electronic transport in high vacuum, the relative intensity of the Raman 'D' and 'G' peak, I_D/I_G, reduces, as shown in Figure 2(c). Because we conduct in-situ transport measurements in the EBL chamber after electron irradiation, it is difficult to conduct Raman spectroscopy experiment immediately after electron irradiation. To reveal the evidence of the reduction of the I_D/I_G after extended vacuum exposure, we irradiated a graphene flake with an electron beam and subsequently collected a Raman spectrum. After keeping the flake in vacuum for 8 hours, we repeated the Raman experiment. While the data in Figure 2(c) and Figure 1(b) are for different devices, we confirmed from multiple graphene flakes that the reduction of the I_D/I_G is a general property which arises from vacuum exposure. Upon electron irradiation on the GFET device, the peak in the resistivity, the Charge-Neutral Point (CNP: also called Dirac Point which is experimentally indicated by a gate voltage where the net carrier concentration is zero) quickly moved from positive to negative gate voltage around -54V (Figure 2(d)). As the irradiated graphene device stays in high vacuum, the CNP slowly moves toward higher gate voltage, but it does not fully recover the value prior to irradiation. Moreover, the resistivity values along the curve were slightly lower after 8h in vacuum. Even though the CNP usually decreases when environmental molecules are removed from a device by annealing or long term vacuum exposure, our graphene devices show an increase of the CNP in Figure 2(d), after exposure to electron beams and vacuum. Because the CNP increases while the irradiated GFET device was exposed to vacuum in Figure 2(d), this shift of the CNP is apparently not caused by desorption of environmental molecules but by other factors. This idea is further supported by the decrease of the I_D/I_G in vacuum (Figure 2(c)). Based on both the CNP shift behavior in vacuum after electron irradiation (Figure 2(d)) and the decrease of I_D/I_G in the Raman spectrum (Figure 2(c)), we argue that the CNP increase in vacuum after electron irradiation is related to desorption of adatoms or molecular fragments, which were generated during the electron irradiation process. This picture is consistent if these adsorbates serve as N-type dopants, because the CNP increases when they desorb from the graphene surface. In vacuum, the CNP does not recover fully. When the device was subsequently exposed to air, the CNP increased by ~60V and approximately recovered its original position in Figure 2(d), which is evidence that the environmental molecules in air are predominantly P-type dopants. As a result, the environmental molecules apparently contribute to recover the original CNP as they readsorb. From the fact that the CNP fully recovered only after interaction of the graphene with environmental molecules in Figure 2(d), we argue that they also contribute to the decrease of the CNP after electron irradiation during which they were partially removed from the graphene surface.

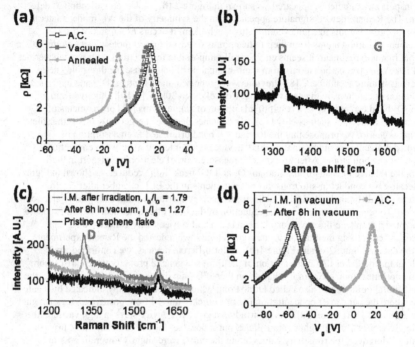

Figure 2. Electrical transport properties and Raman spectrum of the GFET device after electron irradiation of 400uC/cm^2 dose. (a) Resistivity data in ambient condition, vacuum, and after annealing in vacuum before electron irradiation, (b) Raman spectrum of GFET device measured after electron irradiation and in-situ transport measurement, and (c) Raman spectrum of the pristine graphene flake before and after irradiation of electrons with 400uC/cm^2 dosage. The $I_D/I_G = 1.79$ for immediate measurement after electron irradiation and $I_D/I_G = 1.27$ after 8h in vacuum. (d) Resistivity data in vacuum after electron irradiation and in ambient condition after exposing the graphene to air. 'I.M." means Ambient Condition and Immediate Measurement, respectively.

In order to investigate the electrical transport properties after electron irradiation in vacuum further, we irradiated a GFET device with an additional 2000 and 5000uC/cm^2. The ρ_{max} was ~6 kΩ upon first electron irradiation in Figure 2(d) while increasing to 8 and 12kΩ after additional two electron irradiations in Figure 3(a) and (c). In vacuum after the additional electron irradiation, the Dirac points are increased and ρ_{max} is decreased. This is consistent with the result of the 1st electron irradiation in Figure 2(d) and thus further supports our picture of the system.

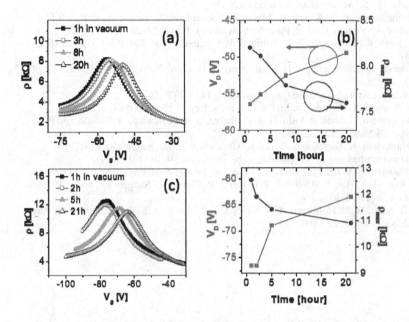

Figure 3. Transport data after additional electron irradiations with accumulated electron dosages of 2000 and 5000 uC/cm^2. (a) Resistivity measured after irradiation with an accumulated electron dosage of 2000uC/cm^2 at times from 1h to 20 h after electron irradiation in vacuum. (b) Dirac point (V_D) and maximum resistivity (ρ_{max}) as a function of time in vacuum (based on data in a). (c) Resistivity for 7000uC/cm^2 and (d) V_D and ρ_{max} data corresponding to (c).

CONCLUSIONS

We have investigated the modified electronic transport properties of graphene FET devices as a result of electron irradiation. We showed that environmental molecules are P-type dopants and contribute to the initial doping behavior of the graphene devices. In addition, we demonstrated that some of the adsorbates are removed when exposed to electron beams. Based on Raman spectroscopy data and Dirac point shifts, we propose that N-type dopants such as fragments of environmental molecules or atoms generated during electron irradiation are also adsorbed on graphene. We consider that a combination of P and N-type dopants contribute to the CNP shift during electron irradiation. In this context, electron irradiation on graphene induces both desorption of environmental molecules and adsorption of fragments of environmental molecules or adatoms, simultaneously.

REFERENCES

1. Schedin, F., Geim, A. K., Morozov, S. V., Hill, E. W., Blake, P., Katsnelson, M. I., Novoselov, K. S, *Nature Materials* **6**, 652 (2007)
2. Subrahmanyam, K. S., Kumar, P., Maitra, U., Govindaraj, A., Hembram, K. P. S. S., Waghmare, U. V., Rao, C. N. R, Proc. Natl. Acad. Sci. **108**, 2674 (2011)
3. J.D. Jones, K.K. Mahajan, W.H. Williams, P.A. Ecton, Y. Mo, J.M. Perez, Carbon 48, 2335 (2010)
4. P. Blakea_ and E. W. Hill, A. H. Castro Neto, A. H. Castro Neto, Appl. Phys. Lett. 91, 063124 (2007)
5. Andrea C. Ferrari, Solid State Communications 143, 47 (2007)
6. J. B. Renucci, R. N. Tyte, and M. Cardona Phys. Rev. B 11, 3885 (1975)
7. H. E. Romero, N. Shen, P. Joshi, H. R. Gutierrez, S. A. Tadigadapa, J. O. Sofo, and P. C. Eklund, ACS Nano 2, 2037 (2008)
8. A. Hashimoto, K. Suenaga, A. Gloter, K. Urita and S. Iijima, nature 430, 870 (2004)
9. D. Teweldebrhan and A. A. Balandin, Appl. Phys. Lett. 94, 013101 (2009)
10. I. Childres, L. A. Jauregui, M. Foxe, J. Tian, R. Jalilian Appl. Phys. Lett. 97, 173109 (2010)
11. S. Ryu, M. Y. Han, J. Maultzsch, T. F. Heinz, P. Kim, M. L. Steigerwald, and L. E. Brus, Nano Lett., 8 (12), 4597 (2008)
12. J. D. Jones, W. D. Hoffmann, A. V. Jesseph, C. J. Morris, G. F. Verbeck, and J. M. Perez, Appl. Phys. Lett., 97, 23 (2010).

Mater. Res. Soc. Symp. Proc. Vol. 1549 © 2013 Materials Research Society
DOI: 10.1557/opl.2013.950

Electrical transport of zig-zag and folded graphene nanoribbons

Watheq Elias[1,2] , M. Elliott[1] and C. C. Matthai[1]
[1]School of Physics and Astronomy
Cardiff University, The Parade, Cardiff, UK
CF24 3AA
[2]Dept of Physics, Koya University, Erbil, Iraq.
E-mail: matthai@astro.cf.ac.uk

ABSTRACT

In recent years, there has been much interest in modelling graphene nanoribbons as they have great potential for use in molecular electronics. We have employed the NEGF formalism to determine the conductivity of graphene nanoribbons in various configurations. The electronic structure calculations were performed within the framework of the Extended Huckel Approximation. Both zigzag and armchair nanoribbons have been considered. In addition, we have also computed the transmission and conductance using the non-equilibrium Greens function formalism for these structures. We also investigated the effect of defects by considering a zigzag nanoribbon with six carbon atoms removed. Finally, the effect of embedding boron nitride aromatic molecules in the nanoribbon has been considered. The results of our calculations are compared with that obtained from recent work carried out using tight-binding model Hamiltonians.

INTRODUCTION

Graphene is a two-dimensional sheet of graphite in which the sp^2 hybridization leads to a trigonal planar structure with strong σ bonds connecting the C-atoms with an equilibrium separation, a, of 1.42Å. It is the strength of these bonds that result in such a robust structure. This crystal structure also gives rise to a unique electronic band structure and a whole range of interesting electronic properties [1]. The high electron mobility and good thermal conductivity of graphene makes it a useful and promising material for use in next generation electronic devices. In particular, their mechanical and electronic properties have enhanced the prospect of graphene nanoribbons (GNR) being used as nanowires in miniaturized electronic devices.

GNR, which are sheets of graphene cut to make ribbons, have two basic edge shapes. The corresponding ribbons are termed zigzag nanoribbons (ZNR) or armchair nanoribbons (ANR). The width of a GNR is defined in terms of the number of C-atoms across the ribbon (N). The same number N for both types of ribbons does not mean that the ribbons have the same width. The width is related to N through the relations:

$$W_A = \left[\frac{N+1}{2}\right]a \text{ for the ANR and } W_Z = \left[\sqrt[3]{N} + \frac{1}{\sqrt[3]{}}\right]a \text{ for the ZNR.}$$

ZNR are metallic and therefore could be used as molecular nanowires. There is therefore much interest in studying the electron transport properties of these nanoribbons. Most of the theoretical investigations on nanowire conductance employ a Landauer description of the tunnelling across the nanowire. These are effectively within the framework of the one electron picture in which transport occurs through the nanowire orbitals on a single electronic potential energy surface. Within this description, the nanowire orbitals can be determined through carrying out first principles calculations like density functional theory (DFT) or by using a semi-empirical approach. Whilst the one-electron nanowire energy spectrum obtained by techniques like DFT are accurate with regard to equilibrium properties, they are computationally expensive and it is not clear to what extent they can be used reliably to strongly correlated transport. By contrast semi-empirical Self Consistent Extended Huckel Theory (SC-EHT) calculations have been carried out in the study of electronic structure of isolated molecules, the band offsets and Schottky barriers at semiconductor interface structures and even the optical spectra of adatom adsorbed surfaces [2].

In this paper, we report on the results of our SC-EHT calculations on the electronic structure and density of states of GNR as a function of their width. We have also carried out electron transport calculations on GNR based on the non-equilibrium Greens function (NEGF) formalism. In addition to investigating the dependence of the conductance on the NR width, we have also made attempts to determine the effect of defects in the form of substitutional molecules on the electrical properties. Recently, Ashhadi and Ketabi [3] (AK) examined the electronic transport properties across boron nitride (BN) molecules embedded between two ZNR.

COMPUTATIONAL APPROACH

The SC-EHT method

The SC-EHT calculations were carried out by employing the non-orthogonal Slater-Koster basis set. The basis functions were taken to be the atomic orbitals on each atom and the overlap matrix formed from the overlap of these atomic orbitals, viz

$$S_{\mu\nu} = \int dr \, \varphi_\mu^*(r) \, \varphi_\nu(r) \qquad (1)$$

The Hamiltonian matrix elements were calculated from the EHT ansatz:

$$H_{\mu\mu} = I_\mu \; ; \; H_{\mu\nu} = 0.5 \, K \, (I_\mu + I_\nu) \, S_{\mu\nu} \qquad (2)$$

where the ionization potentials of each orbital for zero excess charge, $I_\mu(0)$, were taken from the tabulated experimental values. The overlaps between different orbitals centred around the same atom were taken to be zero. In addition, the overlaps between orbitals centred around atoms separated by a distance greater than a cut-off distance were also taken as zero. For distances less than the cut-off distance, the overlaps were determined analytically.

Once the eigenvalue equation was solved, the charge density distribution was determined and from this the excess charge on each site obtained by performing a Mulliken population analysis. This excess charge at each site was used to determine the new ionization potentials which are defined through the equation

$$I_\mu(\Delta q) = I_\mu(0) + \alpha \, \Delta q + \beta \, \Delta q^2 \qquad (3)$$

where the parameters α and β are determined by carrying out all-electron DFT calculations for atoms with some excess charge. This process is iterated until the charge density distribution is self-consistent with the ionization potentials.

Conductance calculations

The current I across a molecular wire is given by

$$I = \frac{2e}{h} \int T(E)[f(E-\mu_1) - f(E-\mu_2)]dE \quad (4)$$

where $T(E)$ is the transmission function across an electrode-molecule-electrode system, $f(E)$ is the Fermi function. μ_i are the electro-chemical potentials of the two electrodes at bias voltage V_B such that $\mu_1 - \mu_2 = eV_B$. $T(E)$ is determined by applying the non-equilibrium Greens function (NEGF) formalism with the molecular wave functions obtained from the SC-EHT calculations.

In the transport calculations on the GNR, the ribbon made up both the leads (electrodes) and the molecule nanowire. In considering the ZNR/BN/ZNR structures, the leads were taken to be GNR units.

RESULTS
Electronic energy states in ANR and ZNR

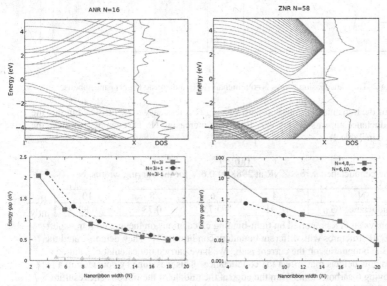

Figure 1: The band structure and density of states for armchair nanoribbons (left) and zigzag nanoribbons (right). The bottom panels show the band gap as a function of N.

The electronic band structure and density of states (DOS) have been calculated for both

43

nanoribbon structures with 3 < N < 28 for ANR and 4 < N < 58 for ZNR. Typical results for both GNR are shown in Figure 1. It can be seen that the main difference in the two sets of results is the presence of the edge states at the Fermi level in the ZNR which are absent in the results for the ANR. The dependence of the band gap as a function of GNR width is also shown in the figure. The results for the band structure are in good agreement with that obtained using a non self-consistent tight-binding approach [4].

Conductance of GNR

The transmission function and the conductance were calculated for N=2, 4, 6, 8 and 10 for both ANR and ZNR structures. As expected from the band structures, the T(E) for the former shows negligible transmission for bias voltages less than 0.5 eV. By contrast the transmission is quite strong for the ZNR structures. In Figure 2 below, the transmission function for N=6 for both structures is plotted.

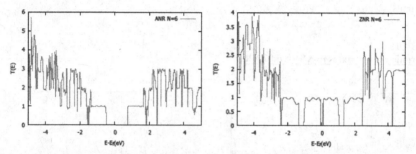

Figure 2: The transmission across N=6 armchair (left) and zigzag (right) nanoribbons.

The variation of the transmission function with varying N is less easy to identify and in fact the calculated conductance appears to vary very little with increasing N. This can be seen from the values tabulated in Table 1.

Table 1
The conductance across ZNR at 298K at 0.6 V for the varying widths, N.

N	4	6	8	10
Conductance (G_0)	0.91	0.73	0.73	0.80

Recently, Takaki and Kobayashi (TK) reported on tight-binding calculations on the quantum transport properties of ZNR structures with different boundary conditions [5]. They demonstrated the importance of the boundaries on the current path. We have carried out calculations aimed at investigating at how defects may affect the conductance. This was done by considering a N=8 ZNR and removing 6 carbon atoms from the edge at the middle of the ribbon. The resulting structure is shown in Figure 3 below.

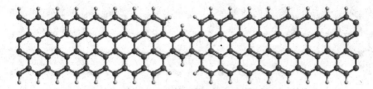

Figure 3: Structure of the N=8 ZNR with six carbon atoms removed.

The transmission function for this structure was calculated and the results are shown in Figure 4. For comparison, the transmission functions calculated for the ZNR for varying N is also given with the same log-scale axis.

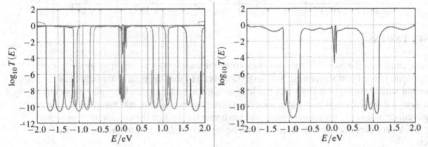

Figure 4 : The transmission function for varying N for the ZNR shown on the left panel. The colours refer to blue (N=4), green (N=6), red (N=8) and cyan (N=10). The right panel shows the transmission for the structure depicted in Figure 3.

The calculated conductance for the structure in Figure 3 is 0.33 G_0 which is less than half that found for the same width perfect ZNR. Since the defect is at the edge of the NR, any change in the conductance may be attributed to edge effects. The magnitude of the reduction lends support to the findings of TK with regard to the current pathway being along the side of the ZNR.

Transport across ZNR/BN/ZNR and ZNR/C/ZNR structures

The electrical transport across boron nitride (BN) aromatic molecules embedded in a ZNR has been studied by AK using a non-self-consistent tight binding method. Their main findings were that there is a dramatic drop in the transmission at all energies even with the introduction of a single BN molecule. We have carried out similar calculations on ZNR/BN/ZNR structures (Figure 5) using our SC-EHT approach. Both I-V curves and conductance were calculated for varying numbers of (BN) nitride units. We find that there is an immediate drop in the conductance when just one BN aromatic molecule is embedded in the ZNR structure. As more units are embedded, the conductance falls exponentially. The results are summarised in Table 2 below.

Figure 5: Structure of ZNR-(BN)$_2$-ZNR.

Table 2
The conductance ZNR/BN/ZNR at 298K at 0.6 V for the varying numbers of (BN) units (M).

M	1	2	3	6
Conductance (G_0)	3×10^{-4}	6×10^{-7}	2×10^{-9}	1×10^{-11}

CONCLUSIONS

We have employed the SC-EHT method to examine the band structure and transmission across the graphene nanoribbon structures. We have demonstrated that the results are in general agreement with those reported by other workers on similar systems. We have also shown that when studying systems in which surface electron states are prominent, it is essential to take charge redistribution effects into account. We have also shown that the conductance across a ZNR is not very dependent on its width but can be severely reduced by the introduction of defects.

ACKNOWLEDGMENTS

Watheq Elias acknowledges support from the Government of Iraq and the Iraqi Cultural Attache in London.

REFERENCES

1. Castro Neto et al, Review of Modern Physics, **81** 109 (2009).
2. Matthai Phil Trans R Soc Lond A , **344**, 137 (1993).
3. Bass and Matthai Phys Rev B, **52**, 4712 (1995).
4. M.Ashhadi and S.A.Ketabi, Physica E 46, 250 (2012).
5. K. Wakabayashi and S. Dutta, Solid State Comm, **152**, 1420 (2012)
6. H Takahi and N Kobayashi, Physica E **43** 711 (2011).

Mater. Res. Soc. Symp. Proc. Vol. 1549 © 2013 Materials Research Society
DOI: 10.1557/opl.2013.1031

Transparent Graphene-Platinum Films for Advancing the Performance of Dye-Sensitized Solar Cells

P. T. Shih[1], R. X. Dong[1], K. C. Ho[1,2,*] and J. J. Lin[1,*]
[1] Institute of Polymer Science and Engineering, National Taiwan University, Taipei 10617, Taiwan.
[2] Department of Chemical Engineering, National Taiwan University, Taipei 10617, Taiwan.

ABSTRACT

Transparent films of platinum nanoparticles on graphene nanohybrids were synthesized in a two-step process. Reduction of homogeneously dispersed Pt precursor and graphene in water and solution coating/annealing afforded thin films with high catalytic performance as counter electrodes in dye-sensitized solar cells (DSSC). The requisite dispersant consisting of poly(oxyethylene)-(POE) segments and cyclic imide functionalities allowed the in-situ reduction of dihydrogen hexachloroplatinate by ethanol and the formation of nanohybrids of graphene-supported Pt nanoparticles at 4.0 nm diameter. Characterizations of polymeric dispersants by Fourier-transform infrared spectroscopy, thermogravimetric analysis, and nanohybrids by transmission electron microscope were performed. After screening various compositions of Pt/graphene, the nanohybrid film at the specific ratio of 5/1 by weight was fabricated into a counter electrode (CE) for DSSC by the solution casting method. The evaluation of cell performance demonstrated the most improved power conversion efficiency of 8.00%. This is significant achievement in comparison with 7.14% for the DSSC with the conventional platinum sputtered CE. Furthermore, the solution casting method allows the preparation of transparent CE films that are suitable for using as rear-illuminated DSSC. The approach was proven to be feasible by measuring the cell efficiency under rear light illumination. The power efficiency up to 7.01%, comparable to 8.00% by a normally front illumination, has been accomplished. In contrast, the rear illumination at merely 2.36% efficiency was obtained for the DSSC with sputtered platinum CE. Analyses of cyclic voltammetry, electrochemical impedance spectra were well correlated to the high efficiency of the performance caused by this nanohybrid film.

INTRODUCTION

Recent advances in synthesizing new nanomaterials allow the controls of nanoparticle size and structural morphology. For example, the manipulation and utilization of carbonbased nanofillers including carbon black, carbon nanotube [1], carbon nanofiber [2], and graphene [3] have been well documented. Among these materials, graphene is unique for the 2D geometric shape and chemical compositions of hexagonally arrayed carbon atoms in single-atom-thick sheet. Owing to its high surface area [4], electrical conductivity [5] and transparency, graphene is potentially useful for fabrication into solar cell devices. However, the sheet-like graphene has the inherent force of aggregating through interlayer piling and difficulty for being dispersed in organic mediums [6,7] in the process of fabricating homogeneous films. In addition, unlike carbon nanotube, graphene is poor in catalytic activity when being used as counter electrodes due to the lacking of edge planes that may facilitate the I_3^- reduction. In view of these inherent properties, the approach of developing new nanohybrids by dispersing the geometrically 2D graphene in combination with platinum nanoparticles (PtNP) has been made. In this study, we report the new strategy of adopting the polymeric surfactant with polar functionalities such as poly(oxyethyelene) segment and imide groups, which serve as noncovalent bonding forces for anchoring the graphene surface and subsequently the PtNP. In the consecutive processes, the dispersed graphene/Pt dispersion could be coated and annealed into thin films for the application of DSSC devices.

EXPERIMENTAL DETAILS

Graphene was obtained from Legend Star International Co., Ltd. Poly(oxyethylene)-diamine with molecular weight (Mw) of 2,000 g/mol (abbreviated POE2000). Monomer, 4,4'-oxydiphthalic anhydride (ODPA, 97% purified by sublimation), Ti (IV) tetra-isopropoxide (TTIP, >98% purity), acetonitrile (ACN), acetylacetone, and isopropyl alcohol (IPA) were purchased from Aldrich Chemical Co. Dihydrogen hexachloroplatinate (H_2PtCl_6, 99.95% purity) was obtained from Alfa Aesar Company.

The requisite dispersant, poly(oxyethylene)-segmented oligo(imide) (POE-imide), was synthesized from an aromatic dianhydride and a poly(oxyethylene)-diamine, preferably 2000 g/mole molecular weight at a proper molar ratio. The product was recovered as a yellowish waxy solid. Nanohybrids of PtNP on graphene nanoplatelets were prepared by an *in-situ* reduction of dihydrogen hexachloroplatinate (H_2PtCl_6) in the presence of POEM dispersant. The graphene was first added into ethanolin a vial and then agitated with a VCX 500 ultrasonicator for an hour at ambient temperature.

The DSSC was assembled with the photoanode and the CE (PtNP/graphene-CE or s-Pt-CE), keeping a distance of 25 mm between them by using Surlyn (Solaronix) as the spacer, and injecting the electrolyte containing 0.6 M DMPII, 0.1 M LiI, 0.05 M I_2, and 0.5 M TBP dissolved in MPN, through a hole between the photoanode and the CE via capillary method.

An EI Electric Co., Ltd. with incident light intensity (100 mW cm^{-2}) was calibrated with a standard Si Cell (PECSI01, Peccell Technologies, Inc.). Photocurrent-voltage curves of the DSSCs were obtained with a potentiostat/galvanostat (PGSTAT 30, Autolab, Eco-Chemie, the Netherlands). The surface morphological view of the PtNP/graphene film was analyzed in SEM (NanoSEM 230, Nova^{TM}). EIS was obtained by the potentiostat/galvanostat mentioned previously. The catalytic ability of the CEs were measured via cyclic voltammetry (CV) was obtained at short-circuit condition. The light source was a class A quality solar simulator (PEC-L11, AM1.5G, Peccell Technologies, Inc.) and the light was focused through a monochromator (Oriel Instrument, model 74100) onto the photovoltaic cell.

DISCUSSION

This polymeric dispersant, POEM, was used for dispersing graphene and supporting the *in situ* reduction of H_2PtCl_6 and stabilizing the generated PtNP. In Figure 1, the representative structure consisting of POE or $-(CH_2CH_2O)_x-$ segments, amidoacids, and aromatic imides $-(CONCO)-$ functionalities, is illustrated. With the addition of H_2PtCl_6 in ethanol/water, the graphene dispersion remained homogeneous and the reduction of Pt salt into PtNP occurred under the mild heating conditions. The generation of homogeneous reduction of PtNP on graphene nanoplatelets was evidenced by the observation of SEM micrographs.

In the loading of POEM to graphene at 10/1 weight ratio, different loadings of H_2PtCl_6 were performed for the reduction. As shown in Figure 2, the SEM images of the homogeneous dispersions derived from PtNP/graphene (w/w) at 1/1, 5/1, and 20/1 are shown and compared to the conventional sputtered Pt (s-Pt) surface. We may apparently observe that 5/1 should be the optimum ratio under which the maximum quantity of PtNPs was reached with the particle size retained the smallest as listed in Table 1. Therefore, the accordingly largest surface area of PtNP was achieved and the highest catalytic ability was exhibited.

The performances of the DSSCs with different PtNP/graphene CEs were evaluated in comparison with that of the cell fabricated by the conventional s-Pt CEs. The current-voltage (I–V) characteristics of the DSSCs were measured under illumination at 100 mW cm^{-2}. The corresponding open-circuit voltage (V_{OC}), short–circuit density (J_{SC}), fill factor (FF) and cell efficiency (η) of the DSSCs are listed in Table 1. The maximal cell efficiency was peaked at 5/1 PtNP/graphene ratio for reaching 8.00%, significantly superior to 7.14% for the conventional s-Pt CE cell. It is explained that the increase in electric conductivity on the electrode surface subsequently accelerates the progress of redox couple regeneration at CE/electrolyte interface where I_3^- may retrieve electrons in a faster manner by the presence of graphene. Besides the factor of graphene, the efficiency increase is well correlated with the finer Pt particle size and the quantity of PtNP in relative to the graphene support.

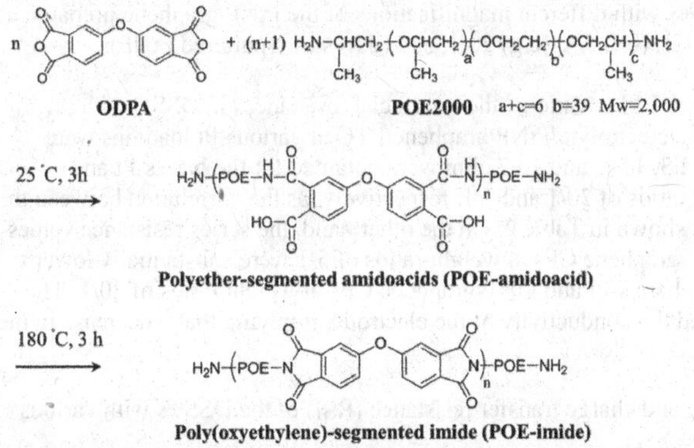

Figure 1. Synthesis of the poly(oxyethylene)-segmented amide-imide (POEM) from dianhydride/diamine via amidation and imidation at high temperature.

Table 1. Photovoltaic parameters of the DSSCs using PtNP/graphene CEs with varying ratios, measured at 100 mW cm^{-2} light intensity in comparison with s-Pt

PtNP: graphene (wt%)	Particle size (nm)	V_{OC} (V)	J_{SC} (mA/cm^2)	FF[a]	η (%)
1 : 1	4.5 ± 0.6	0.722	18.8	0.494	6.71
2 : 1	4.6 ± 0.6	0.691	20.0	0.502	6.94
5 : 1	4.8 ± 0.7	0.710	18.8	0.598	8.00
10 : 1	8.4 ± 1.2	0.731	16.6	0.628	7.62
20 : 1	9.3 ± 1.8	0.685	16.7	0.640	7.32
s-Pt[b]	10–20	0.718	15.5	0.641	7.14

[a] FF = $\eta/(V_{OC} * J_{SC})$ [b] sputter Pt

49

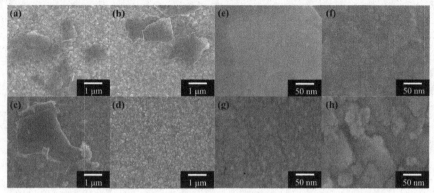

Figure 2. Surface morphologies with different magnifications of the PtNP/graphene prepared at PtNP/graphene (w/w) = (a,e) 1/1, (b,f) 5/1, (c,g) 20/1, and (d,h) s-Pt (sputtered PtNP).

In Figure 3, by employing the EIS technique, the interfacial resistances in DSSC with the configuration of FTO/TiO$_2$/dye/electrolyte/PtNP/graphene/FTO at various Pt loadings were analyzed. The R_{ct1} values of 12.3, 11.9, and 10.7 ohm were obtained for the bare s-Pt and PtNP/graphene CEs at weight ratios of 20/1 and 5/1, respectively, as the correlation between the resistances and cell efficiency shown in Table 2. On the other hand, the series resistance values observed in DSSCs with PtNP/graphene CEs at weight ratios of 5/1 were substantially lower than those from a DSSC using bare s-Pt and PtNP/graphene CEs at weight ratios of 20/1. The presence of graphene enhanced the conductivity of the electrode, implying that a decrease in the value

Table 2. Series resistance (R_s) and charge transfer resistance (R_{ct1}) of the DSSCs with various working electrodes.

PtNP/graphene	R_s ($\Omega\,cm^{-2}$)	R_{ct1} ($\Omega\,cm^{-2}$)	η (%)
5/1	13.5	10.7	8.00
20/1	13.9	11.9	7.32
s-Pt	15.2	12.3	7.14

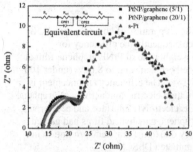

Figure 3. Nyquist plots of EIS spectra of the DSSCs with s-Pt CE and PtNP/graphene CEs with weight ratio of 5/1 and 20/1, measured at 100 mw cm⁻² light intensity.

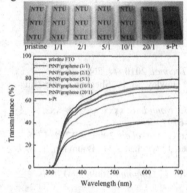

Figure 4. Transmittance of PtNP/graphene CE films at various ratios.

of R_s may accelerate the electron transport within the DSSCs, yielding cells with higher performance.

In addition, regarding the high transparency of the PtNP/graphene film coated on the FTO at counter electrode, UV-Vis spectrophotometer was employed to measure the transmittance of the film comprised of the well-dispersed graphene and various loadings of PtNPs. Figure 4 suggests the differences of CEs in appearance, the optimized one of 5/1 weight ratio, corresponding to 70% transmittance at 550 nm. For the comparison between weight ratio of PtNP/graphene at 5/1 and s-Pt. Efficiency of rear-illuminated solar cell comprised of PtNP/graphene film with the weight ratio of 5/1 is 7.01%, which is merely about 1% behind that of front-illuminated one which of 8.00% under the same conditions, that is even capable of competing with the efficiency of s-Pt (7.14%), let alone s-Pt illuminated from the rear, which of only 2.36%.

CONCLUSIONS

The use of home-made polymeric dispersant allowed the preparation of homogeneous PtNP on graphene films that showed roughness of surface and consequently the efficiency for catalyzing the redox reaction in solar cell. The optimized weight ratio of PtNP/graphene film at 5/1 was fabricated and exhibited the enhancement of the cell efficiency up to 8.00% (under 100 mW cm-2) in comparison to the conventional DSSC of 7.14%. The efficiency enhancement is attributed to the fine distribution of PtNP and the nature of high electric conductivity of graphene that affects the progress of redox couple regeneration at CE/electrolyte interface and accelerates the rate of electron retracing by I3−. In short, the Pt-on-graphene nanohybrid films had gained double advantages in DSSC, enhancement of cell efficiency due to the graphene promotion in electrical conductivity and the possible design of rear-illuminated DSSC.

ACKNOWLEDGMENTS

We acknowledge financial supports from the Ministry of Economic Affairs and National Science Council (NSC) of Taiwan.

REFERENCES

1. Preat, J.; Jacquemin, D.; Perpete, E. A. *Environ. Sci. Technol.* **2010**, *44*, 5666–5671.
2. Zhang, Y.; Zou, G.; Doom, S. K.; Htoon, H.; Stan, L.; Hawley, M. E.; Sheehan, C. J.; Zhu, Y.; Jia, Q. *ACS Nano* **2009**, *3*, 2157–2162.
3. Zhang, Y.; Zhang, L.; Kim, P.; Ge, M.; Li, Z.; Zhou, C. *Nano Lett.* **xxxx**, *xx*, xxx–xxx.
4. Si, Y.; Samulski, E. T. *Chem. Mater.* **2008**, *20*, 6792–6797.
5. Wei, D.; Liu, Y.; Wang, Y.; Zhang, H.; Huang, L.; Yu, G. *Nano Lett.* **2009**, *9*, 1752–1758.
6. Worsley, M. A.; Pauzauskie, P. J.; Olson, T. Y.; Biener, J.; Satcher, J. H.; Baumann, T. F. *J. Am. Chem. Soc.* **2010**, *132*, 14067–14069.
7. Shan, C.; Yang, H.; Han, D.; Zhang, Q.; Ivaska, A.; Niu, L. *Langmuir* **2009**, *25*, 12030–12033.

Mater. Res. Soc. Symp. Proc. Vol. 1549 © 2013 Materials Research Society
DOI: 10.1557/opl.2013.941

Graphyne Oxidation: Insights From a Reactive Molecular Dynamics Investigation

L. D. Machado[1], P. A. S. Autreto[1] and D. S. Galvao[1].

[1]Applied Physics Department, State University of Campinas, 13083-970, Campinas, São Paulo, Brazil.

ABSTRACT

Graphyne is a generic name for a family of carbon allotrope two-dimensional structures where sp^2 (single and double bonds) and sp (triple bonds) hybridized states coexists. They exhibit very interesting electronic and mechanical properties sharing some of the unique graphene characteristics. Similarly to graphene, the graphyne electronic properties can be modified by chemical functionalization, such as; hydrogenation, fluorination and oxidation. Oxidation is of particular interest since it can produce significant structural damages.

In this work we have investigated, through fully atomistic reactive molecular dynamics simulations, the dynamics and structural changes of the oxidation of single-layer graphyne membranes at room temperature. We have considered α, β, and γ-graphyne structures. Our results showed that the oxidation reactions are strongly site dependent and that the sp-hybridized carbon atoms are the preferential sites to chemical attacks. Our results also showed that the effectiveness of the oxidation (estimated from the number of oxygen atoms covalently bonded to carbon atoms) follows the α, β, γ-graphyne structure ordering. These differences can be explained by the fact that for α-graphyne structures the oxidation reactions occur in two steps: first, the oxygen atoms are trapped at the center of the large polygonal rings and then they react with the carbon atoms composing of the triple bonds. The small rings of γ-graphyne structures prevent these reactions to occur. The effectiveness of β-graphyne oxidation is between the α- and γ-graphynes.

INTRODUCTION

Carbon-based materials of reduced dimensionality have shown to exhibit some extraordinary structural, thermal and electronic properties. One example of this is graphene [1], a single-planar layer of sp^2-hybridized carbons that has become one of the hottest topics in materials science today. Due to its unique properties graphene is considered as the basis for a new nanoelectronics [1-3]. However, in its pristine form graphene is a zero bandgap semiconductor, which limits its use in transistor applications [3].

Diverse physical and chemical approaches have been tried to solve this problem [4-6]. Ideally, the gap opening should not compromise other desirable electronic properties, such as, the linear dependence of the energy of the conduction and valence electrons with their momentum, i. e., the main Dirac cone properties. With the approaches mentioned above this has been only partially achieved [1-6].

In part because of this, there is a renewed interest in other possible 2D carbon-based structures, as for example, the graphyne structures (Figure 1) [7,8]. These structures were proposed by Baughman and co-workers [7] and structurally are composed of polygonal rings composed of carbon atoms with the simultaneous existence of sp^2 and sp hybridized states. Graphyne is a generic name for the family of these structures, where many possible forms can exist. In this work we restricted ourselves to investigate the α, β, and γ forms (Figure 1). Graphyne-based nanotubes were also theoretically predicted [9,10] and recently experimentally realized [11]. With relation to the planar structures, although large molecular fragments have been already synthesized, only recently a member of this family (graphdiynes) was successfully synthesized in the form of films [12].

Some of the graphyne structures share some of remarkable graphene properties, such as, the Dirac cone [8,13]. Furthermore, density functional theory (DFT) calculations [14] have shown that the Dirac cone is preserved in the presence heteroatom (B, N, H) dopants, and that the presence of B and N can lead to a small bandgap opening [14]. Oxygen is another heteroatom that can be naturally present in the graphyne synthesis [11,12]. Due to the more pronounced porous structure (Figure 1) it is expected that the presence oxygen atoms can produce more significant electronic and structural changes in graphyne than in graphene, in especial, out-of-plane deformations [15].

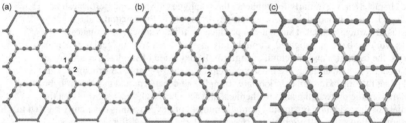

Figure 1. Structure of (a) α-graphyne, (b) β-graphyne and; (c) γ-graphyne. The labeling 1 and 2 refer to non-acetylenic (1) and acetylenic (triple bonds) groups, respectively.

METHODOLOGY

We have carried out fully atomistic molecular dynamics (MD) simulations in order to investigate the structural and dynamical aspects of the atomic oxidation of graphyne membranes. The extensive MD study was carried out using reactive force fields (ReaxFF [16-18]), as implemented in the Large-scale Atomic/Molecular Massively Parallel Simulator (LAMMPS) code [19]. In order to speed up the simulations, we consider a super thermal approach in which O-O recombination was not allowed during the runs. We have used a NVT ensemble and the temperature was controlled using a Nosé-Hoover thermostat, as implemented in LAMMPS code [19]. The simulations were carried out at room temperature (300 K), and the typical time for a complete simulation run was of 500 ps, with time-steps of 0.1 fs.

ReaxFF is a reactive force field developed by van Duin, Goddard III and co-workers for use in MD simulations. It allows simulations of many types of chemical reactions. It is similar to standard non-reactive force fields, like MM3 [16-18], where the system energy is divided into partial energy contributions associated with, amongst others, valence angle bending, bond stretching, and non-bonded van der Waals and Coulomb interactions [16-18]. However, one main difference is that ReaxFF can handle bond formation and dissociation (making/breaking bonds) as a function of bond order values. ReaxFF was parameterized against DFT calculations, being the average deviations between the heats of formation predicted by the ReaxFF and the experiments equal to 2.8 and 2.9 kcal/mol, for non-conjugated and conjugated systems, respectively [16-18].

The process of simulating the oxidation of the graphyne membranes was carried out considering an isolated single-layer graphyne membrane immersed into an atmosphere of atomic oxygen atoms. We considered systems with constant volume and number of atoms. Typical dimensions of the membranes were of 180 Å x 180 Å (~ 6500 atoms) and the atmospheres contained ~6000 atomic oxygen atoms.

RESULTS AND DISCUSSIONS

In Figure 2 we present snapshots from MD simulations after a dynamics period of 0.3 ns. As we can see from this Figure, although all structures contain carbon-carbon triple bonds, the observed level of oxidation is quite different.

These different oxidation dynamics can be better evaluated analyzing the number of covalently bonded oxygen atoms as a function of the time of simulation. These results are presented in Figure 3. As we can see from Figures 2 and 3, the reactivities are graphyne-type and site dependent. The atoms in sp^2-hybridized states only significantly react for the α-graphynes and these site are, as expected, much less reactive and the sp-hybridized ones. The largest site reactivity differences were observed for the β-graphynes. Although a considerable level of oxidation occurs, the oxygen atoms mainly covalently bond to the sp carbons. For γ-graphynes both sites are essentially oxidation-resistant.

In order to try to explain these significant differences we calculated the 3D energy potential maps experienced by an oxygen atom when interacting with the graphyne membranes. These maps provide helpful information about the relative importance of which site can be preferentially attacked.

In Figure 4 we present the results for γ-graphyne, calculated for an oxygen atom placed at 1.2 Å above the membrane basal plane. As we can see from the Figure, for this specific distance value, there are no positions where the C-O interactions are attractive. This helps to explain why γ-graphyne is basically oxidation resistant. Even when the maps are calculated to a height closer to the membranes, the only observed minima were those close to the carbon atoms composing the triple bonds (where the reaction can occur). For most of the cases there are always energy barriers to the oxygen atoms reach the membrane surfaces.

The situation is completely different for the case of α-graphynes. In Figure 5 we present the results for the maps calculated for different basal planar distances. As we can see from the Figure, for the case of h=0.5 Å we observed two energy minima: the deepest one close to the region of the triple bonds and the local one located at the hexagonal rings. Reactions can occur if the oxygen atoms come close to the regions of the global minima. For the case of h=1.5 Å, the energy is positive directly above the triple bonds, which implies that under these conditions there is an energy barrier to approach the sites of the triple bonds. However, for the regions of the center of the hexagon rings, the energies are always negative, although not always these regions correspond to local minima. Since the interactions are always attractive, there are no energy barriers to the oxygen atoms reach these regions. Once these atoms became trapped, they can bounce back and forth close to the membrane surfaces until reactions occur. This greatly increases the chemical reactivity and can explain why α-graphynes can be easily oxidized.

The β-graphyne structures represent an intermediate stage between α- and γ-graphynes. Due do the lack of space we will not discuss this case in details.

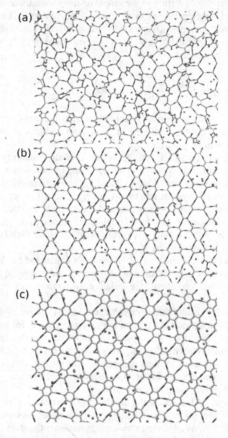

Figure 2. Snapshots from molecular dynamics simulations for oxidized graphyne membranes after a dynamics period of 0.3 ns. Reaction is extensive for the α-graphyne, moderate for the β-graphyne and for the γ-graphyne there is almost no oxidation. Carbon-carbon triple bonds are the preferential targets of the oxygen atoms, leading to the formation of carbon-oxygen double bonds, as well as, epoxy groups.

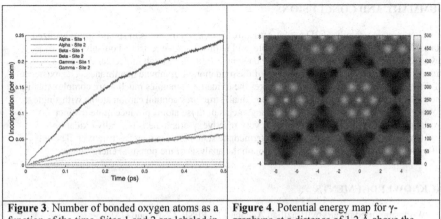

Figure 3. Number of bonded oxygen atoms as a function of the time. Sites 1 and 2 are labeled in Figure 1.

Figure 4. Potential energy map for γ-graphyne at a distance of 1.2 Å above the membrane.

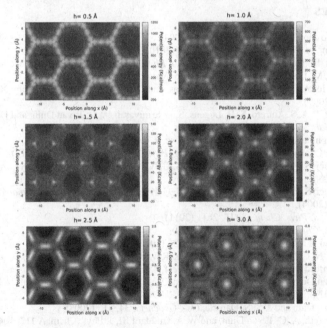

Figure 5. Potential energy maps for α-graphyne. Results for an oxygen atom placed at different distances (h) above the membrane basal plane.

SUMMARY AND CONCLUSIONS

Graphyne is a family of 2D carbon structures composed of atoms in sp and sp^2 hybridized states. Many different structures is possible, and in this work we restricted ourselves to investigate the α, β, and γ-graphynes. We have investigated, through molecular dynamics simulations, the processes and dynamics of the oxidation of graphyne membranes. The existence of different sites that can be oxidized makes the oxidation dynamics much more complex that in the case of graphene oxidation. Also, although all structures contain carbon atoms with single, double and triple bonds, the different arrangements of those atoms produce quite distinct oxidation processes. While α-graphyne is very reactive, γ-graphyne is basically oxidation resistant and β-graphyne represents an intermediate state between α and γ structures. These different reactivities can be explained through the analysis of the potential energy maps.

ACKNOWLEDGEMENTS

Work supported in part by Brazilian Agencies FAPESP, CNPq and CAPES.

REFERENCES

1. K. S. Novoselov *et al.*, Science **306**, 666 (2004).
2. S. H. Cheng *et al.*, Phys. Rev. B **81**, 205435 (2010).
3. F. Withers, M. Duboist, and A. K. Savchenko, *Phys. Rev. B* **82**, 073403 (2010).
4. F. Guinea, M. I. Katsnelson, and A. K. Geim, *Nature Phys.* **6**, 30 (2010).
5. M. Z. S. Flores, P. A. S. Autreto, S. B. Legoas, and D. S. Galvao, *Nanotechnology* **20**, 465704 (2009).
6. R. Paupitz, P. A. S. Autreto, S. B. Legoas, S. G. Srinivasan, A. C. T. van Duin, and D. S. Galvao, *Nanotechnology* **24**, 035706 (2013).
7. R. H. Baughman, H. Eckhardt, and M. Kertesz, *J. Chem. Phys.* **87**, 6687 (1987).
8. D. Malko, C. Neiss, F. Vines, and A. Gorling, *Phys. Rev. Lett.* **108**, 086804 (2012).
9. V. R. Coluci, S. F. Braga, S. B. Legoas, D. S. Galvao, and R. H. Baughman, *Phys. Rev. B* **68**, 035430 (2003).
10. V. R. Coluci, S. F. Braga, S. B. Legoas, D. S. Galvao, and R. H. Baughman, *Nanotechnology* **15**, S142 (2004).
11. C. Li *et al.*, *J. Phys. Chem. C* **115**, 2611 (2011).
12. G. Li *et al.*, *Chem. Commun.* **46**, 3256 (2010).
13. B. Kim and H. Choi, *Phys. Rev. B* **86**, 115435 (2012).
14. D. Malko, N. C. Neiss, and A. Gorling, Phys. Rev. B **86**, 0454434 (2012).
15. S. W. Cranford and M. J. Buehler, *Nanoscale* **4**, 4587 (2012).
16. A. C. T. van Duin, S. Dasgupta, F. Lorant, and W. A. Goddard III, *J. Phys. Chem. A* **105**, 9396 (2001).
17. A. C. T. van Duin and J. S. S. Damste, *Org. Geochem.* **34**, 515 (2003).
18. K. Chenoweth, A. C. T. van Duin, and W. A. Goddard III, J. Phys. Chem. A **112**, 1040 (2008).
19. S. Plimpton, *J. Comp. Phys.* **117**, 1 (1995). http://lammps.sandia.gov/

Mater. Res. Soc. Symp. Proc. Vol. 1549 © 2013 Materials Research Society
DOI: 10.1557/opl.2013.608

On the Dynamics of Graphdiyne Hydrogenation

P. A. Autreto, J. M. de Sousa, and D. S. Galvao

Instituto de Física 'Gleb Wataghin', Universidade Estadual de Campinas, 13083-970, Campinas, São Paulo, Brazil.

ABSTRACT

Graphene is a two-dimensional (2D) hexagonal array of carbon atoms in sp^2-hybridized states. Graphene presents unique and exceptional electronic, thermal and mechanical properties. However, in its pristine state graphene is a gapless semiconductor, which poses some limitations to its use in some transistor electronics. Because of this there is a renewed interest in other possible two-dimensional carbon-based structures similar to graphene. Examples of this are graphynes and graphdiynes, which are two-dimensional structures, composed of carbon atoms in sp^2 and sp-hybridized states. Graphdiynes (benzenoid rings connecting two acetylenic groups) were recently synthesized and they can be intrinsically nonzero gap systems. These systems can be easily hydrogenated and the amount of hydrogenation can be used to tune the band gap value. In this work we have investigated, through fully atomistic molecular dynamics simulations with reactive force field (ReaxFF), the structural and dynamics aspects of the hydrogenation mechanisms of graphdiyne membranes. Our results showed that depending on whether the atoms are in the benzenoid rings or as part of the acetylenic groups, the rates of hydrogenation are quite distinct and change in time in a very complex pattern. Initially, the most probable sites to be hydrogenated are the carbon atoms forming the triple bonds, as expected. But as the amount of hydrogenation increases in time this changes and then the carbon atoms forming single bonds become the preferential sites. The formation of correlated domains observed in hydrogenated graphene is no longer observed in the case of graphdiynes. We have also carried out *ab initio* DFT calculations for model structures in order to test the reliability of ReaxFF calculations.

INTRODUCTION

The chemistry of carbon is very rich, the three different hybridization states (sp, sp^2 and sp^3) allow the generation of a large plethora of distinct structures, such as: graphite (sp^2), diamond (sp^3), fullerene (sp^2), carbon nanotubes (sp^2) and more recently, the hottest topic in materials science, graphene (sp^2) [1]. Graphene, a two-dimensional (2D) sheet of sp^2-hybridized carbon atoms exhibits extraordinary thermal, mechanical, and especially electronic properties. Because of these unique properties graphene is considered one of the most promising materials for future electronics [2]. However, in its pristine state, graphene is a gapless semiconductor, which poses some limitations to its use in transistor electronics [2]. This has renewed the interest in other possible carbon-based 2D materials, as for example, the graphyne structures [3,4]. Proposed by Baughman and co-workers in 1987 [4], graphyne is a generic name for a carbon allotrope family of 2D structures, where benzenoid rings are connected by acetylenic groups (Figure 1), with the coexistence of sp and sp^2 hybridized carbon atoms. These structures share some of graphene unique properties [3], with the advantage of some of them are non-zero electronic bandgap systems [5]. Similar to graphene that can be the structural basis to create carbon nanotubes, graphyne nanotubes are also possible [6,7,8].

In spite of the existence of molecular fragments, only recently a member of this family was successfully synthesized in the form of films: graphdiyne [8]. Graphdiyne (Figure 1) possesses a network composed of two acetylenic ($-C\equiv C-\equiv C-$) groups connecting benzenoid rings. It is expected that the presence of the acetylenic groups can create structures with extraordinary properties, such as: high third-order nonlinear optical susceptibility, thermal resistance, high conductivity, and through-sheet transport of ions [9,10]. Also, similarly to graphene [11], graphdiyne hydrogenation can be used to smartly tune its electronic band gap [10]. With the exception of a recent work on the selective diffusion properties of hydrogen (H_2) on graphdiyne [12], most of the theoretical works addressing the hydrogenation of graphdiyne structures are based on ideally perfect models and the dynamics of the hydrogenation processes remains to be fully investigated. In this work we have investigated, using fully reactive molecular dynamics, the structural and dynamical aspects of the atomic hydrogenation mechanisms of graphdiyne membranes.

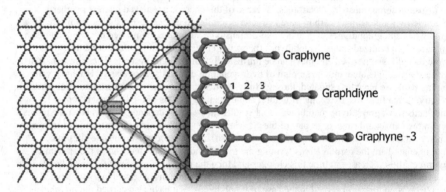

Figure 1. Structural models for the graphyne structures. See text for discussions.

METHODOLOGY

We have carried out fully atomistic molecular dynamics (MD) simulations in order to investigate the structural and dynamical aspects of the atomic hydrogenation of graphdiyne membranes.

The extensive MD study was carried out using reactive force fields (ReaxFF [13-15]), as implemented in the Large-scale atomic/Molecular Massively Parallel Simulator (LAMMPS) code [16]. In order to speed up the simulations, we considered a superthermal approach in which H-H recombination was not allowed during the runs. The Langevin thermostat, as implemented in LAMMPS code [16], was used and the typical time for a complete simulation run was of 500 ps, with time-steps of 0.1 fs.

The process of simulating the hydrogenation of the graphdiyne membranes was carried out considering an isolated single-layer graphdiyne sheet immersed into an atmosphere of atomic hydrogen atoms (see Figure 2). We considered systems with a constant volume and number of particles. Typical dimensions of the graphdiyne membranes were of 185 Å x 150 Å (~ 6200 carbon atoms) and an atmosphere of ~2500 H atoms.

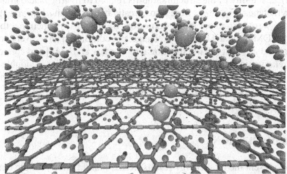

Figure 2. Snapshot from the initial stages of the molecular dynamics simulations of the hydrogenation process of a graphdiyne membrane (stick) immersed into an atmosphere of atomic hydrogen atoms (spheres).

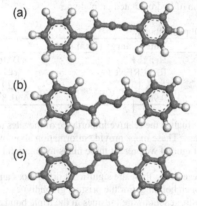

Figure 3. Structural motifs of hydrogenated graphdiynes.

ReaxFF is a reactive force field developed by van Duin, Goddard III and co-workers for use in MD simulations. It allows simulations of many types of chemical reactions. It is similar to standard non-reactive force fields, like MM3 [13-15], where the system energy is divided into partial energy contributions associated with, amongst others, valence angle bending, bond stretching, and non-bonded van der Waals and Coulomb interactions [13-15]. However, one main difference is that ReaxFF can handle bond formation and dissociation (making/breaking bonds) as a function of bond order values. ReaxFF was parameterized against DFT calculations, being the average deviations between the heats of formation predicted by the ReaxFF and the experiments equal to 2.8 and 2.9 kcal/mol, for non-conjugated and conjugated systems, respectively [13-15].

The geometrical stability of the structural motifs related to hydrogenated graphdiynes, the so-called diphenylbutanes (Figure 3), were also investigated using an *ab initio* approach. Total energy calculations were carried out in the framework of the density functional theory (DFT), as implemented in the DMol[3] code [17, 18]. Exchange and correlation terms were considered within the generalized gradient (GGA) functional by Perdew *et al.* [18] and core electrons were treated in a non-relativistic all electron implementation of the potential. A double numerical quality basis set with polarization function (DNP) was considered, with a real-space cutoff of 3.7 Å. The tolerances for energy, gradient and displacement convergence were 0.00027 eV, 0.054 eV^{-1} and 0.005Å, respectively.

RESULTS AND DISCUSSIONS

In Table 1 we present the total energy results (ReaxFF and DMol[3]) for the diphenylbutanes structures shown in Figure 3. As we can see from the table, both methods present the same molecular energy ordering, i.e., the most favorable structure is the one where hydrogen atoms are bonded to the atom positions 1 and 2 (see Figure 1). ReaxFF geometries for these molecules are also in excellent agreement with ab initio (DFT) ones, which corroborate ReaxFF forcefield as a good method for description of hydrogenated graphdiyne.

Table 1: Total energy for diphenylbutanes

Molecules	Total Energy (Kcal/mol)	
	ReaxFF	**DMol**
(a)	REFERENCE	REFERENCE
(b)	1.5×10^{-3}	1.29
(c)	0.98	91.8

One effective way to visualize the relative importance of the sites to be hydrogenated is from the 3D energy potential maps. These maps provide information about which site can be preferentially attacked. In Figure 4 we present these maps for a graphdiyne membrane and its hydrogenated form.

As we can see from these maps, there are significant differences among the possible sites in terms of preferential hydrogen bonding. For the pristine graphdiynes (Figure 4a), the most "attractive" sites for the hydrogenations are the ones in the triple bonds, which are the chemically expected results. However, a single hydrogen atom attached to one of these groups (Figure 4b) is enough to significantly alter the maps.

The hydrogenation processes can produce several and different H-bond configurations. For one hydrogen atom, as can be inferred from the maps of Figure 4, the site 3 (see Figure 1) is the most favorable site to be attacked. For two hydrogen atoms, sites 3 and 2 are the most probable ones. This is a direct consequence of the first atom breaking one of the pi-bonds, thus favoring the hydrogen bonding of its neighboring atoms.

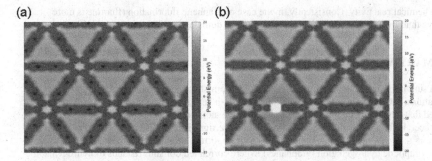

Figure 4. Projected potential energy maps (ReaxFF results) for the potential experienced by a hydrogen atom placed at a distance of 1.5 Å above the basal graphdiyne plane. (a) graphdiyne and; (b) its hydrogenated form.

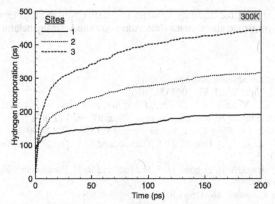

Figure 5. Incorporation of H atoms in time, for the sites indicated in Figure 1. Results for a simulation at 300 K. See text for discussions.

As the number of hydrogen bonded atoms increases in time this pattern changes in a complex way. This can be better visualized following the number of atoms bonded to the different type of sites in time, as presented in Figure 5. As we can see from the Figure, there are two distinct regimes and the percentages of different sites with incorporated hydrogen atoms change in time, although even when the system tends to hydrogen saturation the ordered of the most bonded sites is preserved (3, 2 and 1, respectively). Another important result is that, in contrast with was reported to the case of graphene hydrogenation [11], we did not observe the formation of correlated domains (islands of hydrogenated carbons). Perhaps, this can be a consequence of the porous graphdiyne structure, which allows larger out-of-plane deformations (in comparison to graphene) and, consequently, an increase in the curvature and an increased local chemical reactivity. Consistently, in the case of graphene fluorination (fluorine is more reactive than hydrogen atoms) the formation of these domains are also suppressed [19].

SUMMARY AND CONCLUSIONS

In summary, we have investigated, using fully atomistic reactive molecular dynamics simulations, the dynamics of hydrogenation of single-layer graphdiyne membranes. Our results showed that depending on the atom type (whether the atoms are in the benzenoid rings or as part of the acetylenic groups) the rate of hydrogenation are significantly different. Also, these rates change in time in a very complex pattern. Another important result is that, in contrast with the case of graphene hydrogenation (graphanes), where correlated domains (islands of hydrogenated carbons) were observed, these domains are not formed in hydrogenated graphdiynes.

ACKNOWLEDGEMENTS

The authors wish to thank the Brazilian Agencies FAPESP, CNPq and CAPES for partial funding of this work. We would also to thank Prof. Adri von Duin for many helpful discussions.

REFERENCES

1. K. S. Novoselov *et al.*, *Science* **306**, 666 (2004).
2. S. H. Cheng *et al.*, *Phys. Rev. B* **81**, 205435 (2010).
3. D. Malko, C. Neiss, F. Vines, and A. Gorling, *Phys. Rev. Lett.* **108**, 086804 (2012).
4. R. Baughman, H. Eckhardt, M. Kertesz, *J. Chem. Phys.* **87**, 6687 (1987).
5. Q. Peng, W. Ji, and S. De, *Phys. Chem. Chem. Phys.* **14**, 13385 (2012).
6. V. R. Coluci, S. F. Braga, S. B. Legoas, D. S. Galvao, and R. H. Baughman, *Phys. Rev. B* **68**, 035430 (2003).
7. V. R. Coluci, S. F. Braga, S. B. Legoas, D. S. Galvao, and R. H. Baughman, *Nanotechnology* **15**, S142 (2004).
8. G. Li *et al.*, *Chem. Commun.* **46**, 3256 (2010).
9. G. Luo *et al.*, *Phys. Rev. B* **84**, 075439 (2011).
10. G. M. Psofogiannakis and G. E. Froudakis, *J. Phys. Chem. C* **116**, 19211 (2012).
11. M. Z. S. Flores, P. A. S. Autreto, S. B. Legoas, and D. S. Galvao, *Nanotechnology* **20**, 465704 (2009)
12. S. W. Cranford and M. J. Buelher, *Nanoscale* **4**, 4587 (2012).
13. A. C. T. van Duin, S. Dasgupta, F. Lorant, and W. A. Goddard III, *J. Phys. Chem. A* **105**, 9396 (2001).
14. A. C. T. van Duin and J. S. S. Damste, *Org. Geochem.* **34**, 515 (2003).
15. K. Chenoweth, A. C. T. van Duin, and W. A. Goddard III, *J. Phys. Chem. A* **112**, 1040 (2008).
16. S. Plimpton, J. Comp. Phys. **117**, 1 (1995). http://lammps.sandia.gov/.
17. B. Delly, *J. Chem. Phys.* **92**, 508 (1990).
18. .J. Perdew, K. Burke, and M. Ernzerhof, *Phys. Rev. Lett.* **77**, 3865 (1996).
19. R. Paupitz *et al.*, *Nanotechnology* **24**, 035706 (2013).

Mater. Res. Soc. Symp. Proc. Vol. 1549 © 2013 Materials Research Society
DOI: 10.1557/opl.2013.710

Atomic and Electronic Structure of Multilayer Graphene on a Monolayer Hexagonal Boron Nitride

Celal Yelgel[1] and Gyaneshwar P. Srivastava[1]
[1]School of Physics, University of Exeter, Stocker Road, Exeter, EX4 4QL, U.K.

ABSTRACT

The atomic and electronic structures of multilayer graphene on a monolayer boron nitride (MLBN) have been investigated by using the pseudopotential method and the local density approximation (LDA) of the density functional theory (DFT). We show that the LDA energy band gap can be tuned in the range 41-278 meV for a multilayer graphene by using MLBN as a substrate. The dispersion of the π/π^* bands slightly away from the $\mathbf{K}$ point is linear with the electron speed of 0.9×10^6 and 0.93×10^6 for graphene (MLG)/MLBN and ABA trilayer graphene (TLG)/MLBN systems, respectively. This behaviour becomes quadratic with a relative effective mass of 0.0021 for the bilayer graphene (BLG)/MLBN system. The calculated binding energies are in the range of 10-43 meV per C atom.

INTRODUCTION

Recently, graphene has received tremendous attention as a material of choice in the nano-electronics research area [1, 2]. To design a multifunctional electronic device, graphene needs to have a suitable substrate such as SiO_2 or hexagonal boron nitride (h-BN) [3, 4]. Dean *et al.* [5] have successfully synthesised and characterised high-quality exfoliated mono- and bilayer graphene devices on single-crystal h-BN substrates. Moreover, an effective method to fabricate high-yield two-dimensional h-BN sheets has been developed by using a sonication-centrifugation technique [6]. The h-BN monolayer (MLBN) has a two-dimensional honey-comb structure similar to graphene but contains two chemically inequivalent atomic species per unit cell, making it an insulator with a large band gap. It has been theoretically demonstrated that by depositing graphene on the single or multilayer h-BN a small band gap can be induced [7, 8, 9]. Therefore, it is important to investigate the atomic and electronic structure of multilayer graphene/MLBN systems.

In this study, we have performed the plane wave pseudopotential method, within the density functional scheme, to investigate the equilibrium atomic geometry and electronic structure of multilayer graphene on MLBN. The graphene sheets are found to be weakly adsorbed on the MLBN. We examine the dispersion of the π/π^* bands very close to the $\mathbf{K}$ point. We have shown that it is possible to open a range of band gaps in such a system. The origin of such gap opening is explained by analysing the planer-average electronic charge density difference for the multilayer graphene/MLBN interface along the interface normal.

THEORY

Our investigations are based on the first-principles plane-wave pseudopotential method within the local density approximation (LDA) of the density functional theory. All structures have been modelled by using a repeated slab geometry with a vacuum region of 14 Å. The Perdew-Zunger exchange correlation scheme [10] was considered to treat the electron-electron interactions. The electron-ion interactions were treated by using norm-conserving and fully separable pseudopotentials for carbon, boron, and nitrogen [11]. A plane-wave basis set with a kinetic energy cutoff of 45 Ryd was used to expand the single-particle Kohn-Sham orbitals. Self-consistent solutions of the Kohn-Sham equations were obtained by employing a 36×36×2 **k**-points Monkhorst-Pack set [12] within the supercell Brillouin zone. The geometry was optimised by minimising the Hellmann-Feynman force to smaller than 50 meV/Å. The equilibrium atomic positions were determined by relaxing all atoms in the unit cell except the MLBN. All calculations presented in this study were performed by using our computer code with the method described in Ref. [13].

RESULTS

Before presenting the results of our calculations, we briefly discuss two important issues. The first of these is the use of the LDA for the systems we have studied. The interlayer interaction in these structures is expected to be weak and generally of the van der Waals (vdW) type. Although some recent proposals have been presented to calculate the vdW interactions (the so-called vdW-DFT approach) for graphene-interface systems, both LDA-DFT and vdW-DFT methods are reported to result in very similar interlayer spacings and band structures, despite the LDA-DFT calculations underestimating the adhesion energy [14, 15]. The suitability of the DFT-LDA approach for the energetic and structural properties of graphitic systems is described in previous studies [16, 17]. Therefore, we feel that the DFT-LDA method is appropriate to study the multilayer graphene on the MLBN substrate. The second issue relates to the small lattice mismatch (approximately 1.8%) between graphene and h-BN. Such a lattice mismatch is reported to lead to a moiré superstructure [18, 19], with a periodicity much larger than we can' afford to make first-principles calculations for. We have, however, estimated the range of band gap change of our systems by using the primitive unit cell, viz. the (1×1) cell, but using three lattice constants: 2.45, 2.485 and 2.503 Å which correspond to the equilibrium values for graphene, graphene/BN and h-BN systems, respectively. The maximum difference in the band gap values for these lattice constants is found to be 6 meV.

We first consider monolayer graphene (MLG) interfaced with MLBN. The most stable configuration for this system, with one carbon over B and the other centred above the BN hexagon, is found with the equilibrium distance of 3.22 Å between the MLG and MLBN layers, in agreement with previous studies [20, 7]. The atomic structure is schematically shown in Fig. 1 (a). Our optimisation calculation showed that there is no buckling of MLG. The adsorption energy E_{ads} of this system is obtained by using the following equation

$$E_{ads} = E_{MLG/MLBN} - E_{MLBN} - E_{MLG}, \qquad (1)$$

where the first, second and thrird terms on the right hand side represent the total energies of

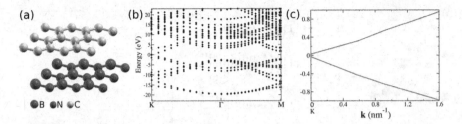

Figure 1: (a) Schematic view of the MLG/MLBN system. (b) Electronic structure of the system. (c) Dispersion curve for the π/π^* bands close to the **K** point.

MLG/MLBN, MLBN and MLG, respectively. The estimated binding energy is 43 meV/C. Figure 1 (b) shows the calculated band structure. We obtained a small band gap of 57 meV, indicating semiconducting behaviour of the system. The behaviour of the π/π^* band close to the **K** point is linear, except for a flattening behaviour in the close vicinity of the **K** point as clarified in Fig. 1 (c). Using the linear part, we estimate the electron speed of 0.9×10^6 m/s. This shows that close to the **K** point the band structure of the MLG/MLBN system is qualitatively identical to graphene.

The atomic structure of the BLG/MLBN system is shown in Fig. 2 (a). In this system, AB-stacked BLG is deposited on MLBN by using the most stable configuration for the MLG/MLBN system. The distance of 3.22 Å was used between the MLBN and the BLG. To optimise the structure, the BN layer was held fixed, and the atoms in BLG were allowed to relax. After the optimisation, the bottom and top layers of the BLG are buckled by 0.01 Å and 0.02 Å, respectively. The binding energy of 27 meV/C is estimated for this system. There is a band gap of 278 meV at the **K** point. The dispersion curve for the π/π^* band shows a quadratic behaviour near the **K** point, as presented in Fig. 2 (c). Our calculations reveal that the Dirac point is shifted and split. We also calculated the electron effective mass of $0.0021m_e$ for this system. This value is almost ten times smaller than the effective mass for BLG. These results suggest that the electronic properties of BLG can be altered using MLBN as a substrate.

As represented in Fig. 3 (a), we modelled the TLG/MLBN system by using the ABA-stacked TLG and its separation from the MLBN of 3.22 Å. Our relaxation calculations reveal that only the C atoms on the MLBN is buckled by 0.001 Å. We found a reduced binding energy of 10 meV/C. Figure 3 (b) shows the band structure of TLG/MLBN. We obtained a band gap of 41 meV for this system. As shown in Fig. 3 (c), the behaviour of the highest and the lowest bands slightly away from the **K** point is mostly linear. Our partial charge density calculation suggests that these bands at the **K** point are derived from the B atom on the bottom layer and the C atom on the bottom layer of TLG lying above the B atom. Using the linear part of the unoccupied band near **K** the electron velocity is calculated as 0.93×10^6 m/s. It is interesting to note that our theoretical value of the electron velocity for the TLG/MLBN almost matches with graphene. We also found that the Dirac point is located at the **K** point but is split.

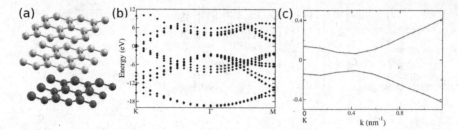

Figure 2: (a) Structure of the BLG/MLBN system. (b) The electronic band structure of the system. (c) The dispersion curves for the inner pair of the π/π^* bands near the **K** point.

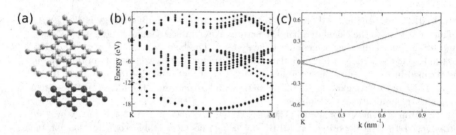

Figure 3: (a) Structure of the ABA-stacked TLG/MLBN system. (b) Band structure of the system. (c) The dispersion curve of the inner pair of the π/π^* bands close to the **K** point.

To investigate the origin of the band gap opening for the MLG/MLBN interface, the planer-average electronic charge density difference along the interface normal is calculated using the following expression

$$\Delta\rho = \rho[\text{MLG/MLBN}] - \rho[\text{MLG}] - \rho[\text{MLBN}], \qquad (2)$$

where the terms on the right hand side represent the charge density of the MLG/MLBN, MLG, and MLBN systems, respectively. We found that there is a re-distribution of the charge density around the graphene sheet as shown in Fig. 4 (a). This leads to the development of a dipole across the graphene sheet. The magnitude of the dipole moment per unit cell in the z direction can be computed as follows:

$$p = -\int \rho(z)z\,\mathrm{d}z + \sum_i Z_i e z_i, \qquad (3)$$

where $\rho(z)$ is the valence electron density integrated over the x-y plane [$\rho(z) \geq 0$ by definition], $Z_i e$ is valence charge on the ith atom in the unit cell, $-e$ is the electronic charge ($e > 0$), and z_i is the z-coordinate of the ith atom. We estimated the electric dipole moment

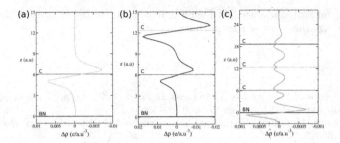

Figure 4: The planer-average electronic charge density difference $\Delta\rho$ along the interface normal direction for (a) MLG/MLBN, (b) BLG/MLBN, and (c) TLG/MLBN.

of magnitude 0.019 Debye for the MLG/MLBN system. The corresponding development of the electrostatic potential across a unit cell was numerically calculated using the expression

$$\Delta V = -4\pi e p. \qquad (4)$$

Our estimated magnitude of the electrostatic potential is 0.102 eV nm^{-1}. This internal electrostatic potential opens up a small gap of 57 meV at the $\mathbf{K}$ point for the MLG/MLBN system. This can be explained as an internal Stark effect.

Similarly, our calculations show that there is also a charge re-distribution around the bottom and top layer of the BLG (see Fig. 4 (b)) and leads to a band gap of 278 meV with the calculated magnitude of the electrostatic potential of 0.083 eV nm^{-1}. There are both theoretical and experimental calculations that clearly point out that the application of an external electric field perpendicular the layers can induce a band gap for BLG [21, 22]. For the TLG/MLBN system, we calculated the energy band gap opening of 41 meV is traced to arise from a charge accumulation between the MLBN and the layer of TLG closest to MLBN, as seen in Fig. 4 (c). We found a dipole moment of magnitude 0.03 Debye for this system and estimated the electrostatic potential of 0.104 eV nm^{-1}.

SUMMARY

In summary, by performing DFT-LDA calculations we show that the electronic band gap and the electron effective mass in multilayer graphene can be modified and tuned by using MLBN as a substrate. The binding energy per C atom is found to gradually decrease as the number of graphene layers increases. Our calculations also suggest an oscillatory behaviour in the band gap opening: 57 meV for MLG/MLBN, 278 meV for BLG/MLBN, and 41 meV for ABA-TLG/MLBN. The opening of the band gap is due to the interaction between graphene and its substrate. We have demonstrated that the MLBN substrate alters the electronic structure of multilayer graphene. The dispersions of the highest valence and the lowest conduction bands are linear for MLG/MLBN and TLG/MLBN, but show significant quadratic behaviour for BLG/MLBN.

CY is grateful for financial support from The Republic of Turkey Ministry of National Education through University of Rize in Turkey.

REFERENCES

[1] K. S. Novoselov, A. K. Geim, S. V. Morozov, D. Jiang, Y. Zhang, S. V. Dubonos, I. V. Grigorieva, and A. A. Firsov, *Science* **306**, 666 (2004).

[2] P. Neugebauer, M. Orlita, C. Faugeras, A. L. Barra, and M. Potemski, *Phys. Rev. Lett.* **103**, 136403 (2009).

[3] J. Hofrichter, B. N. Szafranek, M. Otto, T. J. Echtermeyer, M. Baus, A. Majerus, V. Geringer, M. Ramsteiner, and H. Kurz, *Nano Lett.* **10**, 36 (2010).

[4] D. Usachov, V. K. Adamchuk, D. Haberer, A. Grüneis, H. Sachdev, A. B. Preobrajenski, C. Laubschat, and D. V. Vyalikh, *Phys. Rev. B* **82**, 075415 (2010).

[5] C. R. Dean, A. F. Young, P. Cadden-Zimansky, L. Wang, H. Ren, K. Watanabe, T. Taniguchi, P. Kim, J. Hone, and K. L. Shepard, *Nature Physics* **7**, 693 (2011).

[6] C. R. Dean, A. F. Young, I. Meric, C. Lee, L. Wang, S. Sorgenfrei, K. Watanabe, T. Taniguchi, P. Kim, K. L. Shepard, and J. Hone, *Nat. Nanotech.* **5**, 722 (2010).

[7] G. Giovannetti, P. A. Khomyakov, G. Brocks, P. J. Kelly, and J. van den Brink, *Phys. Rev. B* **76**, 073103 (2007).

[8] J. Slawinska, I. Zasada, and Z. Klusek, *Phys. Rev. B* **81**, 155433 (2010).

[9] Y. Fan, M. Zhao, Z. Wang, X. Zhang, and H. Zhang, *Appl. Phys. Lett.* **98**, 083103 (2011).

[10] Perdew J. P. and Zunger A., *Phys. Rev. B* **23**, 5048 (1981).

[11] X. Gonze, R. Stumpf, and M. Scheffler, *Phys. Rev. B* **44**, 8503 (1991).

[12] H. J. Monkhorst and J. D. Pack, *Phys. Rev. B* **13**, 5189 (1976).

[13] G. P. Srivastava, *Theoretical Modelling of Semiconductor Surfaces*, (World Scientific, Singapore, 1999).

[14] Y. Fan, M. Zhao, Z. Wang, X. Zhang, and H. Zhang, *Appl. Phys. Lett.* **98**, 083103 (2011).

[15] B. Sachs, T. O. Wehling, M. I. Katsnelson, and A. I. Lichtenstein, *Phys. Rev. B* **84**, 195414 (2011).

[16] B. Carsten, L. Predrag, D. Rabie, C. Johann, G. Timm, A. Nicolae, C. Vasile, B. Radovan, T. N. Alpha, B. Stefan, Z. Jörg, and M. Thomas, *Phys. Rev. Lett.* **107**, 036101 (2011).

[17] G. R. Victor, L. Wei, Z. Egbert, S. Matthias, and T. Alexandre, *Phys. Rev. Lett.* **108**, 146103 (2012).

[18] M. Kindermann, B. Uchoa, and D. L. Miller, *Phys. Rev. B* **86**, 115415 (2012).

[19] C. Ortix, L. Yang, and J. Brink, *Phys. Rev. B* **86**, 081405(R) (2012).

[20] C. Yelgel and G. P. Srivastava, *Appl. Surf. Sci.* **258**, 8338 (2012).

[21] Y. Zhang, T. Tang, C. Girit, Z. Hao, M. C. Martin, A. Zett, M. F. Crommie, Y. R. Shen, and F. Wang, *Nature* **459**, 820 (2009).

[22] E. V. Castro, K. S. Novoselov, S. V. Morozov, N. M. R. Peres, J. M. B. Lopes dos Santos, J. Nilsson, F. Guinea, A. K. Geim, and A. H. Castro Neto, *J. Phys.: Condens. Matter* **22**, 175503 (2010).

Other 2D-Layered Materials

Mater. Res. Soc. Symp. Proc. Vol. 1549 © 2013 Materials Research Society
DOI: 10.1557/opl.2013.792

Ambipolar transport in MoS₂ based electric double layer transistors

Jianting Ye[1], Yijin Zhang[1], and Yoshihiro Iwasa[1]
[1]Quantum Phase electronics center and department of Applied Physics, The University of Tokyo, Tokyo, Japan

ABSTRACT

Making field effect transistors (FETs) on thin flake of single crystal isolated from layered materials was pioneered by the success of graphene. To overcome the difficulties of the zero band gap in graphene electronics, we report the fabrication of an electric double layer (EDL) transistor, a variant of FET, based on another layered material, MoS₂. Using strong carrier tunability found in EDL coupled by ion movement, MoS₂ transistor displayed an unambiguously ambipolar operation in addition to its commonly observed n-type transport. A high on/off ratio $>10^4$, large "ON" state conductivity of ~mS, and a high reachable $n_{2D} \sim 1\times10^{14}$ cm^{-2} confirmed the high performance transistor operation being important for application. The high-density carriers of both holes and electrons can drive the MoS₂ channel to metallic states indicating that new electronic phases could be accessed using the protocol established in making EDL gated transistors on layered materials.

INTRODUCTION

Layered materials led by graphene are now attracting great interests by the success of the Scotch-tape method, which offers a simple and effective process for field effect transistor (FET) device fabrication based on thin flakes isolated from their bulk layered crystals [1,2]. Although graphene-FETs exhibit high performance such as high mobility up to ~ 10^4 Vs/cm^2 and ambipolar operation [1], the gapless nature in its Dirac band structure prevents device application in real world. To overcome the difficulty in graphene-FETs to realize "OFF" state in device operation, semiconducting transition-metal dichalcogenides, represented by molybdenum disulfide MoS₂, are regarded as promising candidates for FET device with its gapped nature for creating "OFF" state [3]. The first demonstration of MoS₂-FET with SiO₂ as the gate dielectrics following the same manner as graphene only showed a poor modulation in conductivity [2], while the later attempt with HfO₂, one of the high-k materials, as the gate dielectrics realized a clear switching operation with ON/OFF ratio up to 10^8 [4]. The straight forward interpretation of this improvement may be the enhancement of charge accumulation capability with improved gate efficiency from SiO₂ to HfO₂. Although a practical FET device can be realized using HfO₂ dielectrics, a significant nature of ambipolarity with highly symmetric operation as demonstrated in graphene-FET is still absent [1]. A further improvement of the gate efficiency is necessary to reach such a ambipolar operation in MoS₂. We demonstrated an ambipolar FET operation in MoS₂ by enhancing the gate efficiency using electric double layer transistor (EDLT) with ionic liquid as the gate dielectrics. EDLT is a proved device structure for significant enhancement of transport [5].

EXPERIMENT

MoS$_2$ single crystal was grown using chemical vapor transport (CVT) method [6]. Firstly, stoichiometric amounts of pure Mo and S powders were mixed, sealed into a quartz ampoule under high vacuum, and repeatedly heated in a Muffle furnace at 700 °C and 1100 °C to make high-quality polycrystalline MoS$_2$ powders. Secondly, polycrystalline powders were mixed with pure I$_2$, sealed into another ampoule, and heated in two-zone furnace to carry out CVT.

Grown MoS$_2$ single crystals were mechanically exfoliated to isolate thin flakes and transferred onto sapphire of SiO$_2$/Si++ substrates by the Scotch-tape method [1,2]. Proper thin flakes with atomically flat surface were selected under optical microscope, followed by the micro-fabrication process of electron-beam lithography, metal evaporation, and lift-off. A droplet of ionic liquid, DEME-TFSI (*N,N*-diethyl-*N*-methyl-*N*-(2-methoxyethyl) ammonium bis (trifluoromethylsulfonyl)imide) was placed to cover channel surface and the gate electrode, forming an EDLT device configuration.

All measurements were carried out inside a physical property measurement system (Quantum Design, Inc.) under high vacuum (~ 10^{-5} Torr) and at temperatures below 220 K to suppress possible chemical reactions.

RESULT

Figure 1 shows a typical transfer curve of MoS$_2$-EDLT with a full cycle scan of gate voltage V_{GS} and output curves under various V_{GS}. In addition to commonly known *n*-type operation of MoS$_2$ [2,4], additional hole current appeared under V_{GS} < -2 V resulted in an ambipolar operation [7]. The ambipolarity also exhibits itself in output curves by the upturn at V_{DS} > 2 V under V_{GS} < 0.3 V.

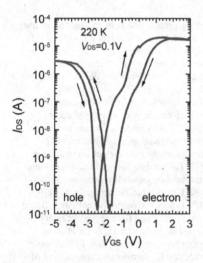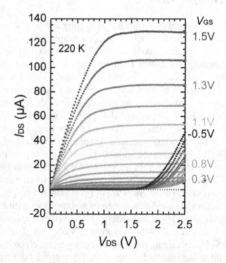

Figure 1. Transfer curve and output curves of MoS_2-EDLT. The ambipolar operation was established by the first observation of *p*-type operation in addition to the conventional *n*-type operation.

Additionally, Hall effect measurement revealed that the charge carrier density reached ~ 10^{14} /cm^2 both for electron and hole doping and their mobility was ~ 50 and ~ 100 for electron and hole, respectively. The higher mobility of holes indicated that the newly revealed hole conduction might be more useful for FET devices than conventional electron conduction. Due to the large carrier density realized only by EDLT, the MoS_2 channel surface was driven into metallic nature, surprisingly, both under electron and hole doping as shown in figure 2 [7].

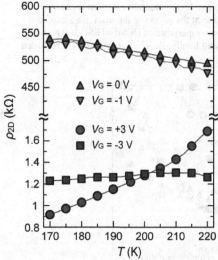

Figure 2. Temperature dependence of channel conductivity measured with four-terminal configuration. Either electron or hole accumulation can drive insulator-metal transition.

Field induced *p-n* junction

As both types of carriers can be selectively induced by changing polarity of gate voltage, correctively sophisticated bias condition will form a kind of *p-n* junction inside channel surface by adjusting V_{GS} and V_{GS} to the different polarity, for instance, a condition of $V_{GS} > 0 > V_{GD}$ will induce electrons at the source electrode and holes at the drain electrode simultaneously. The formation of *p-n* junction can be electrically traced through the output curve under a fixed V_{GS} and varying V_{DS} (and thus, V_{GD} is also changing). As shown in the inset of figure 3, output curve shows a second upturn at high V_{DS} indicating an appearance of a *p-n* junction.

Cooling-while-gating technique

As is obvious from the formation mechanism, the field-induced p-n junction is highly sensitive to both V_{GS} and V_{DS}. Thus its I-V characteristics cannot be investigated. To overcome this difficulty, we investigated a cool-while-gating method taking advantage of the liquid nature of the gate dielectrics. Conventionally, the information of gate voltage will directly affect carrier profile in the channel. In EDLTs, on the other hand, these twos are mediated by ion reformation driven by the potential variation between gate electrode and channel surface. This additional step in the mechanism will, however, add a new possibility of keeping charge profile fixed by freezing the gate dielectrics suppressing ion motion at low temperature. We cooled down the device with various condition of biases as indicated by allows in the inset of figure 3 to compare the I-V characteristics with the evolution of p-n junction. Experimentally, we first measured output curve at 220 K and stopped scanning V_{DS} at one of the allows in the inset, then cooled down the device to 180 K passing the glass transition temperature of DEME-TFSI, 200 K, to freeze gate dielectrics while the device was biased, and finally, I-V characteristics were recorded by scanning V_{DS} at 180 K.

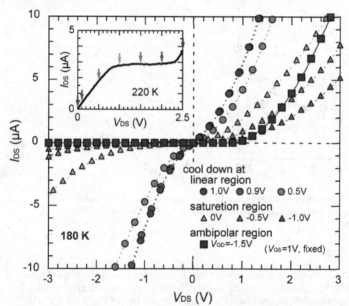

Figure 3. Low temperature I-V characteristics of various junctions controlled by the combination of V_{GS} and V_{DS}.

When only electrons were induced at the channel surface, I-V characteristics were symmetric to voltage (circles in figure 3). After the pinch-off occurred and channel surface was non-uniformly doped, anisotropic I-V characteristics showed up (triangles in figure 3). By further increase of V_{DS} at 220 K forming a *p-n* junction, *I-V* characteristics changed into a highly rectifying curve as shown by squares in figure 3. Please notice that positive and negative V_{DS} correspond to forward and reverse bias, respectively, in a traditional *p-n* junction from doped semiconductors. This result is a proof for a field-induced *p-n* junction.

CONCLUSIONS

We fabricated EDLT device using MoS_2 for channel materials and observed *p*-type operation for the first time. Collaborated with the conventional *n*-type operation of MoS_2-FETs, high performance ambipolar FET was realized with high ON/OFF ratios and symmetric operations. The significant improvement in the device performance is the consequence of intruding EDLT device structure which can realize charge carrier accumulation up to ~ $10^{14}/cm^2$. Such a high carrier density further drove an ambipolar insulator-metal transition of MoS_2. Recently, deeper investigation in electron side of MoS_2-EDLT has demonstrated the field induced superconductivity [8]. Similar investigation for hole side may be reveal an exotic phenomena such as ambipolar superconductivity.

Additionally, we utilized the liquid nature of EDLT gate dielectrics by the combination of cool-while-gating technique, to stabilize carrier accumulation in the channel. This technique enables us to measure *I-V* characteristics of the field-induced *p-n* junctions showing a rectifying operation similar to the conventional *p-n* junctions.

ACKNOWLEDGMENTS

We thank Prof. S. Ishiwata for experimental support in crystal growth. Prof. T. Takenobu and Dr. Y. Yomogida are highly appreciated for experimental help and deep discussions. This work was supported by a Grant-in-Aid for Scientific Research (S) (21224009), the FIRST Program from JSPS, and SICORP from JST.

REFERENCES

1. K. S. Novoselov, A. K. Geim, S. V. Morozov, D. Jiang, Y. Zhang, s. V. Dubonos, I. V. Grigorieva, and A. A. Firsov, *Science* **306**, 666 (2005)
2. K. S. Novoselov, D. Jiang, F. Schedin, T. J. Booth, V. V. Khotkevich, S. V. Morozov, and A. K .Geim, *Proc. Natl. Acad. Scie. USA* **102**, 10451 (2005)
3. J. A. Wilson, and A. D. Yoffe, *Adv. Phys.* **18**, 193 (1969)
4. B. Radisavlijevic, A. Radenovic, J. Brivio, V. Giacometti, and A. Kis, *Nat. Nanotechnol.* **6**, 147 (2011)
5. H. Shimotani, H. Asanuma, and Y. Iwasa, *Jpn, J. Appl. Phys.* **46**, 3613 (2007)
6. A. A. Al-Hilli, and B. J. Evans, *J. Cryst. Growth* **15**, 93 (1972) / S. H. Mahalawy, and B. L. Evans, *Phys. Stat. Sol. (b)* **79**,713 (1977)
7. Y. J. Zhang, J. T. Ye, Y. Matsuhashi, and Y. Iwasa, *Nano Lett.* **12**, 1136 (2012)

8. J. T. Ye, Y. J. Zhang, R. Akashi, M. S. Bahramy, R. Arita, and Y. Iwasa *Science*, **338**, 1193 (2012)

Mater. Res. Soc. Symp. Proc. Vol. 1549 © 2013 Materials Research Society
DOI: 10.1557/opl.2013.709

Field-induced superconductivity in MoS$_2$

Y. J. Zhang[1], J. T. Ye[1], and Y. Iwasa[1,2]

[1] Quantum-Phase Electronics Center and Department of Applied Physics, The University of
Tokyo, 7-3-1 Hongo, Bunkyo-ku, Tokyo 113-8656, Japan
[2] CERG, RIKEN, Hirosawa 2-1, Wako 351-0198, Japan

ABSTRACT

We fabricated MoS$_2$ transistor adopting electric double layer (EDL) as gate dielectric. So far, EDL has realized p-type conducting MoS$_2$ in addition to well-known n-type conduction showing ambipolar operation. In our study, field-effect superconducting transition of MoS$_2$ was realized with maximum T_C around 10 K. This T_C is the highest not only within MoS$_2$ compounds but also among whole TMDs. The highest T_C discovered in this study lies in the carrier density region much smaller than chemically investigated region. Such compounds with small doping level have never been successfully synthesized by chemical method. Furthermore, by combining HfO$_2$ (typical high-k material for FETs) gating with EDL gating, continuous control of carrier density, and thus quantum phase, was demonstrated. As a result, we successfully obtained the phase diagram of MoS$_2$. Interestingly, the T_C exhibits strong carrier density dependence, showing dome-shaped superconducting phase. Superconducting dome in other materials than cuprates has been reported only a few times in doped 2D semiconductors. Since FET charge accumulation is basically two dimensional, our result implies the existence of common mechanism for superconducting dome in 2D band insulators.

INTRODUCTION

Recent years saw a great success in making two-dimensional electronic systems starting from the discovery of new ways to prepare ultra-thin single crystals pioneered by the research of graphene, a monolayer of graphite, using a simple method of cleaving bulk graphite with Scotch tapes [1–3]. The success of graphene immediately raised the possibility of applying similar technique to other layered materials. The early examples include atomic layers of boron nitride, transition-metal dichalcogenides, and complex oxides like layered high T_c cuprate. The prepared nano-sheets appeared to be stable 2D crystals under ambient conditions exhibiting high crystal quality on a macroscopic scale [2]. Since many layered material were well-studied with varieties of properties of charge, orbital, and spin in their bulk form, isolating them into atomically thin nano-sheets provide new opportunities to study them in a different paradigm [4].

Apart from the development of making nano-sheets, independently, people have been dreamed for a long-time to modulate quantum phase transition using electric field effect. One of the most well-known ideas is to manipulate the superconducting transition using field effect generated by a transistor. Early theory [5–7] and experimental effect [8] started even in the dawn of superconductivity. Later, the introduction of strong gate dielectrics (ferroelectric) significantly improved the control on transition temperature T_c from 10^{-4} K [8] to a level of ~10 K [9,10]. Complete switching ON/OFF of a superconducting states was demonstrated on oxide interfaces of SrTiO$_3$ using the solid gating or recently developed new device structure using electric double layer (EDL) gate dielectrics [11,12]. As we will elaborate afterwards, the charge accumulation ability of EDL gated transistor (EDLT) was significantly improved after using ionic liquid, a molten organic salt under room temperature, leading to a charge accumulation to the order of ~10^{14} cm^{-2} to induce superconductivity in layered materials [13,14].

Combining the advantage of the making ultra-thin flakes and highly efficient EDL-gated transistors, new effort was made to accumulate high-density carriers using EDLT devices structures on high-quality channel surface on the cleaved nano-sheets. Form fundamentals of making such a device, we describe below how this combination had enabled recent success in inducing superconductivity [13-15], creating new transport properties [16,17], and modulating magnetism [18] on these nano-sheets. These results represented a rapidly growing field with emerging opportunities for future researches.

EXPERIMENT

Nanosheets isolated from layered bulk MoS$_2$ were fabricated by exfoliation of bulk crystals using an adhesive tape. The produced thin flakes were subsequently transferred onto a substrate covered by solid dielectrics. Usually, we used 300 nm SiO$_2$ [13,16]. The high-k dielectric for higher solid gate efficiency was recently introduced [14]. The thickness of the flakes was determined by analyzing optical micrographs in either transmission or reflection mode [17]. A 20 nm thick nanosheet (measured by AFM) was subsequently patterned into a Hall bar configuration (Fig. 1a, top panel) using conventional micro-fabrication techniques (electron beam lithography, electron-beam evaporation, and lift-off). The relatively thick nanosheet enabled us to work with relatively higher mobility [19]. The accumulation of large carrier density at the surface is enabled by the use of ionic liquids, which are composed only by organic ions with a much higher ion concentration to be reached [16], resulting in the reduction of the thickness of the electric double layer together with the concomitant increase in geometrical capacitance.

Transport measurements were performed by applying a small AC voltage between source and drain electrode, and using a lock-in amplifier to measure the transverse, V_{xx}, and longitudinal, V_{xy}, voltage drops in Hall bar devices as a function of gate bias voltage V_{LG} applied on the side gate electrode to move the ions in the ionic liquid (Fig 1a, bottom panel) to induce a quantum phase transition such as field-induced superconductivity. A gate bias voltage V_{BG} was also applied to the substrate though the HfO$_2$ gate dielectrics to finely tune the electronic states.

RESULT

Field-induced superconductivity in MoS$_2$

Using the double gating technique described above, we are able to continuously tune the carrier density with a high-k dielectric: HfO$_2$ even though the movement of ionic liquid was frozen at the low temperature (Fig. 1a). This is important for the fine-tuning of the superconducting states and metal-insulator transition after setting a channel state using liquid gating. As shown in Fig. 1b, after charging carrier onto the channel surface at 220 K with different liquid gate biases V_{LG}, the four-terminal sheet resistance R_s was measured as a function of temperature T when the device was cooled down to 2K. Transport at all gate biases $V_{LG} < 1V$ shows negative temperature derivative of R_s reduction ratio $\dfrac{dR_s}{dT}$. Increase of $\dfrac{dR_s}{dT}$ with the increase of gate bias V_{LG} indicates the gradual formation of degenerate carriers and enhanced mobility against the localization of carriers towards the low temperature. The channel surface shows metallic transport at gate bias $V_{LG} \geq 1V$ showing positive $\frac{dR_s}{dT}$. Enhancement of metallicity continues with further increase of V_{LG}. Superconductivity emerges at $V_{LG} = 4V$. The transition temperature T_c shows clear V_{LG} dependence until V_{LG} reaches 6 V.

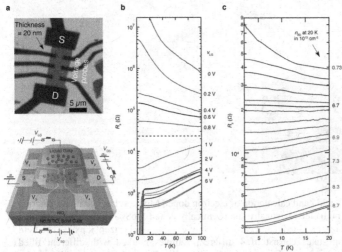

Figure 1. a, Optical micrograph of a typical MoS$_2$ device (d = 20 nm, AFM) under transmission light illumination. **b**, Temperature dependence of channel sheet resistance R_s at different V_{LG} gate bias from 0 to 6 V. **c**, Temperature dependence of channel sheet resistance R_s at V_{LG} = 1 V and different V_{BG} showing metal insulator transition at $n_{2D} = 8.7 \times 10^{12}$ cm^{-2}. For each R_s, we marked the corresponding n_{2D} measured by Hall effect at 20K.

For the metallic states at $V_{LG} = 1V$, we switched on the solid back gate to study the metal insulator transition (MIT) by a precisely control of the transport with 15 different n_{2D} measured by the Hall effect at 20 K ($7.3 \times 10^{11} < n_{2D} < 8.7 \times 10^{12}$ cm^{-2}) as shown Fig. 1c. A transition between the insulating and metallic transport was clearly separated by a critical $R_c = 21.7$ kΩ at $n_c = 6.7 \times 10^{12}$ cm^{-2}, close to the quantum resistance h/e^2, in consistent with the MIT found in other 2D electron systems [20]. A Hall mobility of $\mu_H \sim 240$ cm^2/Vs at MIT is comparable to the bulk value found in MoS$_2$ single crystals [21].

81

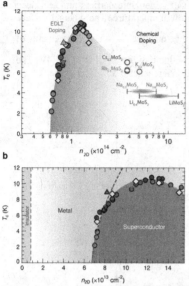

a

EDLT
Doping

Chemical
Doping

$Cs_{0.1}MoS_2$ K_xMoS_2
$Rb_{0.3}MoS_2$ O

$Na_{0.4}MoS_2$ $Na_{0.x}MoS_2$

$Li_{0.x}MoS_2$ $LiMoS_2$

n_{2D} (×10^{14} cm^{-2})

b

Insulator

Metal

Superconductor

n_{2D} (×10^{13} cm^{-2})

Figure 2. Phase diagram and band calculation of electron doped MoS_2. **a**, Unified phase diagram of superconductivity of both electrostatically and chemically doped MoS_2 as a function of doping concentration x (upper horizontal axis), and carrier density n_{2D} (bottom horizontal axis). The field induced superconducting states consist of 4 different samples marked with different filled symbols. All filled circles correspond to one sample, where the same color corresponding to the superconducting states at a fixed V_{LG} and different V_{BG}. The open circles show the T_c of chemically intercalated MoS_2 with different alkali metal dopants. The solid bars denote the range of doping showing the same T_c. The structure of all intercalated compounds remains to be $2H$-type within the indicated carrier density region. **b**, Phase diagram of superconductivity of electrostatically doped MoS_2. The colors indicate different ground state covering the different n_{2D} at low temperature (red, insulating, gray, metallic, and green, superconducting). The dashed line corresponding to a fitting according to $T \propto (n_{2D} - n_0)^{z\bar{v}}$, where $z\bar{v} = 0.6$ is in consistent with 3D-XY model.

Dome-like superconducting phase diagram

For field-induced superconductivity, we expect the carriers are accumulated in a very narrow range close to the surface of the sample. We can compare the phase diagrams thus formed and compare them with chemically doped bulk counterparts. For carrier accumulated by field effect, as estimated by Thomas-Fermi screening length under the present $n_{2D} \sim 10^{14}$ cm^{-1}, we assumed that the carriers are populated in a half of one unit cell, 6.15 Å (a monolayer). Fig. 2b shows a superconducting phase diagram of MoS_2 as function of n_{2D}. It is worth noting that the present n_{2D} can be regarded as the upper limit in the estimation of n_{2D} when we unify our phase diagram with alkali metal intercalated $2H$-MoS_2 in Fig. 2a (26, 27). With the increase of n_{2D} after entering the metallic phase, the superconductivity sharply appears at $n_{2D} = 6.8 \times 10^{13}$ cm^{-2} ($V_{LG} =$

4 V). The initial increase of critical temperature T_c with n_{2D} is saturated and reaching the maximum $T_c = 10.8$ K at $n_{2D} = 1.2 \times 10^{14}$ cm^{-2}, and decreased at larger n_{2D} resulting in a dome-like phase diagram. The unique point found in field induced superconductivity with respect to the bulk phenomena is the onset of superconducting phase, a critical behavior near the quantum critical point could be well fitted a by a relationship of $T \propto (n_{2D} - n_0)^{z\bar{v}}$, where $z\bar{v} = 0.6$ in similar manner as shown in LaAlO$_3$/SrTiO$_3$ interface [12]. This quantum critical phenomena is also consistent with $z\bar{v} = 0.5 \sim 0.6$ found in boron-doped diamond where a comparable T_c was observed [22].

Compared with alkali metal doped compound [23,24], the field-induced phase showed enhanced T_c ($\sim 40\%$ higher than maximum T_c found in Cs$_x$MoS$_2$) at a much lower n_{2D} as shown in Fig. 2a. This temperature is also well above that of NbSe$_2$ with $T_c = 7$K, the original maximum T_c in transition metal dichalcogenides. As shown in Fig. 2b, in the onset of superconductivity ($n_{2D} = 6.8 \times 10^{13}$ cm^{-2}), sharp increase of T_c with the increase of n_{2D} can be descried by conventional BCS theory until reaching $T_c = 10.8$ K at the dome peak. At higher n_{2D}, evolution of T_c versus n_{2D} is opposite to the BCS theory. Phonon softening as a function of n_{2D} is usually regarded as the cause of this anomalous [25], but fails to be the origin in MoS$_2$ since its structural transition occurs at a order of magnitude higher n_{2D} [23]. For small n_{2D} superconductors, other scenarios: large changes in electron-phonon interactions after a function of n_{2D} due to screening may occur as proposed for bulk SrTiO$_3$ [26], or alternative bosons such as plasmons can possibly mediate superconductivity [6]. Similarities between present low onset n_{2D} and small DOS and optimal doping superconductivity in MoS$_2$ is similar to other superconducting interfaces [11–13] and doped band insulator [26,27], which might suggest a possible unified superconducting phase diagram for doped band insulator at dilute carriers.

CONCLUSIONS

From the results we obtained, ionic liquid gating appears to be an effective and reliable technique to accumulate very large amounts of carriers at liquid/solid interface on layered materials. Although our investigations have mainly focused on basic aspects of the electronic properties, the technique has the potential to have a much broader impact, both to disclose new phenomena of fundamental interest and for possible future applications. The present observation of a dome-like phase diagram, an enhanced T_c, and the electrostatic method to induce them may also pave a promising path to novel electronic devices such as superconducting transistors by harvesting the relatively high T_c of 10 K in MoS$_2$. The dome-like phase diagram might host field-controllable competing phase parameters rich for new electronic functionalities, which would certainly be an emerging unexplored paradigm for the future researches.

ACKNOWLEDGMENTS

We thank Y. Takada and A. Bianconi for fruitful discussions and M. Nakano and Y. Kasahara for experimental help. This work was supported by Grant-in-Aid for Scientific Research (S) (No. 21224009) from Japan and Strategic International Collaborative Research Program (SICORP), Japan Science and Technology Agency. M.S.B, R.A and Y.I. were supported by the Japan Society for the Promotion of Science (JSPS) through its Funding Program for World-Leading Innovative R&D on Science and Technology (FIRST Program).

REFERENCES

[1] K. S. Novoselov, A. K. Geim, S. V. Morozov, D. Jiang, Y. Zhang, S. V. Dubonos, I. V. Grigorieva, and A. A. Firsov, Science **306**, 666 (2004).

[2] K. S. Novoselov, D. Jiang, F. Schedin, T. J. Booth, V. V. Khotkevich, S. V. Morozov, and A. K. Geim, PNAS **102**, 10451 (2005).

[3] A. K. Geim and K. S. Novoselov, Nature Materials **6**, 183 (2007).

[4] Q. H. Wang, K. Kalantar-Zadeh, A. Kis, J. N. Coleman, and M. S. Strano, Nat. Nanotechnol. **7**, 699 (2012).

[5] D. Allender, J. Bray, and J. Bardeen, Phys. Rev. B **7**, 1020 (1973).

[6] Y. Takada, J. Phys. Soc. Jpn. **45**, 786 (1978).

[7] S. Brazovskii, N. Kirova, and V. Yakovenko, Solid State Communications **55**, 187 (1985).

[8] R. E. Glover, III and M. D. Sherrill, Phys. Rev. Lett. **5**, 248 (1960).

[9] C. H. Ahn, J. M. Triscone, and J. Mannhart, Nature **424**, 1015 (2003).

[10] C. H. Ahn, A. Bhattacharya, M. Di Ventra, J. N. Eckstein, C. D. Frisbie, M. E. Gershenson, A. M. Goldman, I. H. Inoue, J. Mannhart, A. J. Millis, A. F. Morpurgo, D. Natelson, and J.-M. Triscone, Rev. Mod. Phys. **78**, 1185 (2006).

[11] K. Ueno, S. Nakamura, H. Shimotani, A. Ohtomo, N. Kimura, T. Nojima, H. Aoki, Y. Iwasa, and M. Kawasaki, Nat. Mater. **7**, 855 (2008).

[12] A. D. Caviglia, S. Gariglio, N. Reyren, D. Jaccard, T. Schneider, M. Gabay, S. Thiel, G. Hammerl, J. Mannhart, and J.-M. Triscone, Nature **456**, 624 (2008).

[13] J. T. Ye, S. Inoue, K. Kobayashi, Y. Kasahara, H. T. Yuan, H. Shimotani, and Y. Iwasa, Nature Materials **9**, 125 (2010).

[14] J. T. Ye, Y. J. Zhang, R. Akashi, M. S. Bahramy, R. Arita, and Y. Iwasa, Science **338**, 1193 (2012).

[15] J. T. Ye, S. Inoue, K. Kobayashi, Y. Kasahara, H. T. Yuan, H. Shimotani, and Y. Iwasa, Physica C-Superconductivity and Its Applications **470**, S682 (2010).

[16] J. Ye, M. F. Craciun, M. Koshino, S. Russo, S. Inoue, H. Yuan, H. Shimotani, A. F. Morpurgo, and Y. Iwasa, PNAS **108**, 13002 (2011).

[17] Y. Zhang, J. Ye, Y. Matsuhashi, and Y. Iwasa, Nano Lett. **12**, 1136 (2012).

[18] J. G. Checkelsky, J. Ye, Y. Onose, Y. Iwasa, and Y. Tokura, Nat. Phys. **8**, 729 (2012).

[19] W. Bao, X. Cai, D. Kim, K. Sridhara, and M. S. Fuhrer, Applied Physics Letters **102**, 042104 (2013).

[20] E. Abrahams, S. V. Kravchenko, and M. P. Sarachik, Rev. Mod. Phys. **73**, 251 (2001).

[21] R. Fivaz and E. Mooser, Phys. Rev. **163**, 743 (1967).

[22] T. Klein, P. Achatz, J. Kacmarcik, C. Marcenat, F. Gustafsson, J. Marcus, E. Bustarret, J. Pernot, F. Omnes, B. E. Sernelius, C. Persson, A. Ferreira da Silva, and C. Cytermann, Phys. Rev. B **75**, 165313 (2007).

[23] R. B. Somoano, V. Hadek, and A. Rembaum, The Journal of Chemical Physics **58**, 697 (1973).

[24] J. A. Woollam and R. B. Somoano, Materials Science and Engineering **31**, 289 (1977).

[25] H. R. Shanks, Solid State Communications **15**, 753 (1974).

[26] J. F. Schooley, W. R. Hosler, E. Ambler, J. H. Becker, M. L. Cohen, and C. S. Koonce, Phys. Rev. Lett. **14**, 305 (1965).

[27] Y. Taguchi, A. Kitora, and Y. Iwasa, Phys. Rev. Lett. **97**, 107001 (2006).

Mater. Res. Soc. Symp. Proc. Vol. 1549 © 2013 Materials Research Society
DOI: 10.1557/opl.2013.859

Electron microscopic characterization of multi-layer boron nitride nanosheets

Muhammad Sajjad and Peter Feng[*]

Department of Physics, College of Natural Sciences, University of Puerto Rico, P.O. Box 70377, San Juan, PR/USA   00936-8377

[*]Corresponding author: e-mail: p.feng@upr.edu

Keywords: laser ablation, layered structure, nanoscale

Abstract

We report on the direct synthesis of multi-layer boron nitride nanosheets (BNNSs) and their electron microscopic characterization. The synthesis process is carried out by irradiating hexagonal boron nitride (*h*-BN) target using short laser pulses. Scanning electron microscopy showed large area ($\approx 50 \times 50$ μm^2) flat layers of BNNSs transparent to the electron beam. Low magnification transmission electron microscope (TEM) is used to characterize different areas of nanosheets. TEM revealed that each individual nanosheet is composed of several layers. High resolution TEM (HRTEM) measurements confirmed the layered structure. HRTEM analysis of the edge of a nanosheet showed 10 layers from which we obtained the thickness (3.3nm) of an individual nanosheet. Selected area electron diffraction pattern indicated polycrystalline structure of nanosheets. Raman spectroscopy clearly identified E_{2g} vibrational mode related to *h*-BN.

Introduction

2-dimentional (2D) crystalline materials have many unique properties such as high chemical and mechanical strength and impermeability to gases etc., thus making them very attractive for many applications. The 2010 Noble prize in physics for the work on graphene emphasizes the importance of this new paradigm of 2D crystalline materials research. Geim et al. [1] and many other researches [2,3] identified a 2-dimentional (2D) nanosheet of carbon atoms arranged in honeycomb network named as graphene that has exceptional properties which are not observed in 3-dimentional (diamond like carbon) and 1-dimentional (nanotubes and nanowires) form of carbon. In addition to graphene, other 2D materials such as BNNSs can also be mechanically exfoliated [4] and they can remain stable in air. Layered BN is a structural analogue of graphene in which alternating B and N atoms substitute for C atoms. Because of its appearance in the bulk form, it is often called "white graphene" [5]. BNNSs also exhibit superior properties (greater chemical stability, tunable electronic and sensing properties) compared to bulk BN and have potential applications in electronic devices and in novel functional materials for nanoscale engineering [6,7]. Progress in the growth of BNNSs is available by exfoliation method; but the limited size of samples is still a problem. Recently, chemical vapor deposition techniques (CVD) [8-10] is applied to grow large samples, however presence of chemical precursor in the samples remains a challenge. In our previous report, we applied CO_2-pulsed plasma laser deposition (CO_2-PLD) technique and yielded large amount of pure BNNSs [11].

In this paper, we report on the electron microscopic characterization of crystalline BNNSs synthesized on silicon substrate. The nanosheets were surveyed using Raman spectroscopy and examined using scanning electron microscopy (SEM), transmission electron microscopy (TEM) and high resolution TEM (HRTEM) microscopy. SEM images were recorded using a JEOL JSM-7500F, also equipped with a transmission electron detector (TED). TEM data were recorded using a Carl Zeiss LEO 922 EFTEM and a JEOL JEM-2200FS high resolution TEM (HRTEM).

Experimental Conditions

The synthesis process is carried out by irradiating a h-BN target (2″ diameter, 0.125″ thick, 99.99% purity, B/N ratio $\approx$ 1, density: 1.94 g/cc) using a CO_2 pulsed laser deposition technique (CO_2-PLD: wavelength 10.6 μm, pulse width of 1-5 μs, repetition rate: 5 Hz, pulse energy 5 J, and power density 2×10^8 W/cm^2 per pulse). The advantage of short pulse deposition is to reduce the stress in the sheets that prevents the layers to break unless large area, flat BNNSs finally obtained. The pressure in the deposition chamber was kept at 2.66 x 10^3 Pa. A highly polished, mirror shined silicon (Si) wafer used as a substrate was mounted onto substrate holder 3 cm away from the target. Prior to the deposition, Si substrate was cleaned with acetone and rinsed with deionized water to remove the surface impurities. The laser beam was focused with a 30 cm focal length ZnSe lens onto the target with an incident angle of 45° relative to the surface of rotating target (speed of circa 200 rpm). The purpose to use long focal length ZnSe lens is to effectively control the laser-produced plasma beams. Shift of lens position in front of the target can be adjusted to control power density and size of plasma beam. A heater and a thermocouple were used to obtain and monitor the desired substrate temperature (450 °C). The duration for the deposition was kept 15 min.

Results and Discussion

Raman spectroscopy is used to analyze the crystalline structure of nanosheets. Fig. 2a shows a typical Raman spectrum recorded from BNNSs. An intense central line was observed at 1365 cm^{-1} indicating Raman active E_{2g} vibrational mode related to h-BN [12,13]. No other peaks associated with impurity elements were detected indicates that hexagonal crystalline structure is dominant in BNNSs.

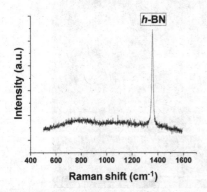

Figure 1. Raman spectroscopy of BNNSs.

Electron microscopic images of BNNSs are presented in Fig. 2. SEM image (Fig. 2a) is indicating several nanosheets highly flat and transparent to the electron beam. Corresponding TEM and HRTEM images of BNNSs are presented in Fig. 2(b-d). TEM data shows several multilayer nanosheets (Fig. 2b). In order to interpret the structure and number of layers in nanosheets, we selected a piece of an individual nanosheet as shown in Fig. 2b and focused electron beam onto its edge. Three layers can be identified as shown in Fig. 2c. Continuing to magnify the TEM image, the highly ordered layer structure become obvious, where seven layers can be distinguished (Fig. 2d). Such morphology clearly shows that each nanosheet is composed of several layers depends on the thickness of nanosheet. So far, at this stage it is difficult to predict the accurate thickness and actual number of layers in a single nanosheet due to limited resolution of TEM. In order to identify crystalline structure more precisely and to report the correct number of layers, HRTEM measurements were required. This time, HRTEM image (Fig. 2e) collected from single nanosheet shows 10 layers. As the distance between each two consecutive layers is 0.33 nm, the thickness of nanosheet is around 3.3 nm. Selected area electron diffraction (ED) pattern represents complicated pattern of several diffraction rings coming from various over-layer structures of BNNSs representing the polycrystalline structure of nanosheets (Fig. 2f). Meanwhile, with the careful observation, one can distinguish bright dots displaying hexagonal patterns.

The general interpretation for the formation of nanosheets is that; when laser pulse impacts the target surface, it produces plasma plume composed of high energy BN ions and clusters. These ions and clusters moves along the plasma column reached to the surface of substrate where they play important roles such as surface heating and etching of the substrate. Meanwhile, substrate heating provides extra thermal vibrations that cause energy exchange through collisions between the active ions, clusters and atoms of BN. This helped to transform small size of BNNSs that coalescence and transformed in to a finite nanosheet. The detail mechanism can be found in our previous publication [11].

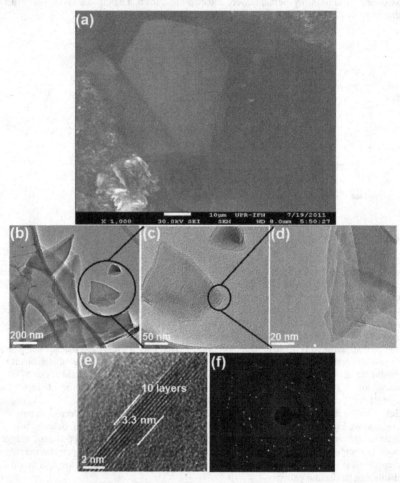

Figure 2. Electron microscopic images of BNNSs, (a) SEM, (b) TEM, (c,d) magnified TEM images of the edge of nanosheet, (e) HRTEM image, and (f) selected area ED pattern. The scalar bars are shown on the image.

Energy dispersive X-ray spectroscopy (EDS) result of BNNSs is presented in Fig. 3. EDS shows that nanosheets are mainly composed of B and N species. A small amount of oxygen is also detected that could possibly come from air to the surface of nanosheets [14,15].

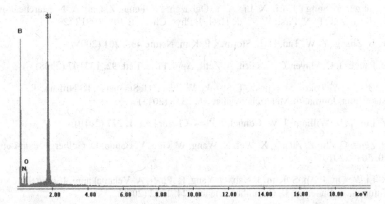

Figure 3. EDS spectrum of BNNSs.

Conclusions

Compared to several other deposition techniques used for the synthesis of nanostructures and thin films, the present pulsed laser plasma beam deposition technique is superior to produce high purity, multi-layer BNNSs. Each individual nanosheet is composed of several layers that can be separated via chemical exfoliation method similar method used for the exfoliation of graphene. Average size of each continuous BNNSs piece is around 50 micrometer squares. The obtained nanosheets are polycrystalline and hexagonal crystalline structure is dominant.

Acknowledgements

This work is partially supported by UPR-DEGI and NASA-EPSCoR seed grants.

References

[1]   A. K. Geim, K. S. Novoselov, Nature Materials **6**, 183 (2007).

[2]   C. Berger, Z. Song , T. Li , X. Li , A. Y. Ogbazghi , R. Feng , Z. Dai , A. N. Marchenkov, E. H. Conrad , P. N. First, W. A. de Heer, J. Phys. Chem. B **108**, 19912 (2004).

[3]   Y. B. Zhang, Y. W. Tan, H. L. Stormer, P. Kim, Nature **438**, 201 (2005).

[4]   D. Pacile, J. C. Mayer, C. O. Girit, A. Zettl, Appl. Phys. Lett. **92**, 133107 (2008).

[5]   H. B. Cho, Y. Tokoi, S. Tanaka, T. Suzuki, W. Jiang, H. Suematsu, K. Niihara, T. Nakayama, Journal of Materials Science **46**, 2318 (2011).

[6]   Y. Lin, T. V. William, J. W. Connel, J. Phys. Chem. Lett. **1**, 277 (2010).

[7]   H. Zeng, C. Zhi, Z. Zhang, X. Wei, X. Wang, W. Guo, Y. Bando, D. Golberg, Nano Lett. **10**, 5049 (2010).

[8]   X. Li,W. Cai, J. An, S. Kim, J. Nah, D. Yang, R. Piner, A. Velamakanni, I. Jung, E. Tutuc, S. K. Banerjee, L. Colombo, Science **324**, 1312 (2009).

[9]   L. Song, L. Ci, H. Lu, P. B. Sorokin, C. Jin, J. Ni, A. G. Kvashnin, D. G. Kvashnin, J. Lou, B. I. Yakobson, P. M. Ajayan, Nano Lett. **10**, 3209 (2010).

[10]   L. Ci, L. Song, C. Jin, D. Jariwala, D. Wu, Y. Li, A. Srivastava, Z. F. Wang, K. Storr, L. Balicas, F. Liu, P. M. Ajayan, Nature Materials **9**, 430 (2010).

[11]   M. Sajjad, M. Ahmadi, M. J-F Guinel, Y. Lin, P. X. Feng J Mater Sci. (2012).

[12]   A. Nag, K. Raidongia, K. P. S. S. Hembram, R. Datta, U. V. Waghmare, C. N. R. Rao ACS Nano **4**, 1539 (2010).

[13]   M. M. Hoffman, G. L. Doll, P. C. Eklund, Phys. Rev. B **30**, 6051 (1984).

[14]   M. Sajjad, X. P. Feng Low temperature synthesis of cubic boron nitride films. Appl. Phys. Lett. **99**, 253109 (2011).

[15]   M. Sajjad, H. X. Zhang, X. Y. Peng, P. X. Feng, Phys. Scr. **83**, 065601 (2011).

Mater. Res. Soc. Symp. Proc. Vol. 1549 © 2013 Materials Research Society
DOI: 10.1557/opl.2013.793

The Hydrogenation Dynamics of h-BN Sheets

Eric Perim[1], Ricardo Paupitz[2], P. A. S. Autreto[1] and D. S. Galvao[1].

[1]Instituto de Física 'Gleb Wataghin', Universidade Estadual de Campinas, 13083-970,
Campinas, São Paulo, Brazil.

[2]Departamento de Física, IGCE, Universidade Estadual Paulista, UNESP, 130506-900, Rio
Claro, SP, Brazil.

ABSTRACT

Hexagonal boron nitride (h-BN), also known as white graphite, is the inorganic analogue of graphite. Single layers of both structures have been already experimentally realized.

In this work we have investigated, through fully atomistic reactive molecular dynamics simulations, the dynamics of hydrogenation of h-BN single-layers membranes.

Our results show that the rate of hydrogenation atoms bonded to the membrane is highly dependent on the temperature and that only at low temperatures there is a preferential bond to boron atoms. Unlike graphanes (hydrogenated graphene), hydrogenated h-BN membranes do not exhibit the formation of correlated domains. Also, the out-of-plane deformations are more pronounced in comparison with the graphene case. After a critical number of incorporated hydrogen atoms the membrane become increasingly defective, lost its two-dimensional character and collapses. The hydrogen radial pair distribution and second-nearest neighbor correlations were also analyzed.

INTRODUCTION

The advent of graphene [1-3] created a new era in materials science. Graphene [1] is a single-layer of sp^2-hybridized hexagonal array of carbon atoms, which exhibits unique electronic, structural and mechanical properties. Because of these properties graphene is considered to be one of the most promising materials for a new electronics [2]. However, in its pristine form graphene is a zero bandgap semiconductor, which limits its use in some kind of transistor applications [2-3].

Much effort has been devoted to try to find a route to open, in a controlled way, a gap in the graphene band structure. The most common strategies explore physical and chemical methods, including chemical modifications through hydrogenation [4-6] and fluorination [7-8]. Hydrogenated graphenes are called graphanes [5] and fluorinated ones, fluorographenes [7-8]. The success of these approaches to solve the graphene bandgap problem has been only partially achieved [4-8].

In part because of this there is a renewed interest in other graphene-like structures, as for example, in the hexagonal boron nitride (h-BN) [9-10]. h-BN, also known as inorganic graphite or white graphite, is the structural analogue of graphite, presenting the same morphology of pilled up honeycomb sheets. Even the interlayer and bond distances are almost identical, the only important structural difference being the layer stacking order.

However, the synthesis of BN structures are, in general, difficult and only recently h-BN monolayers (the equivalent to graphenes) have been obtained [9-12].

In this work we have investigated the dynamics of hydrogenation of h-BN single-layer (what would be the inorganic equivalent of graphanes, the hydrogenated graphenes). We have analyzed the effects of the changes of local hybridizations (from sp^2 to sp^3 ones) produced by the hydrogen bonding. We have also investigated the dynamics of N-H and N-B bond formation during the hydrogenation processes. The similarities and differences with relation to graphene were also addressed.

METHODOLOGY

The dynamics of the hydrogenation of h-BN single-layer membranes was investigated through fully atomistic molecular dynamics (MD) simulations using the reactive force field ReaxFF [13-14], as implemented in the Large-scale Atomic/Molecular Massively Parallel Simulator (LAMMPS) package [15].

ReaxFF [13] is a force field that uses a bond length/bond order relationship, updating the bond order values at each interaction. It is similar to standard non-reactive force fields, like MM3 [13-14], where the system energy is divided into partial energy contributions associated with, amongst others, valence angle bonding and bond stretching, as well as, non-bonded van der Waals and Coulombic interactions. The main ReaxFF advantages over other non-reactive methods are that it can handle bond formation and dissociation (making/breaking bonds) as a function of the bond order values. ReaxFF was parameterized against density functional theory (DFT) methods, being the average deviation between the heats of formation predicted by theory (ReaxFF) and the experiments equal to 2.8 and 2.9 Kcal/mol, for non-conjugated and conjugated systems, respectively [13-14].

This method allows the simulation of many types of chemical reactions ant it has been proven to be very effective on the study of dynamical aspects of a great number of carbon-based structures, including carbon nanotubes [16] and graphene-like systems [6,8].

The MD simulations were carried out under a NVT ensemble, at 50 K, temperature controlled via a Nosé-Hoover thermostat [15]. Typical integration time-steps of 0.1 fs and total simulation time of about 1 ns for each MD run were used.

In order to simulate the hydrogenation processes we considered h-BN single-layer membranes with typical dimensions of 100 Å x 100 Å, immersed into an atmosphere of about 16000 hydrogen atoms (mimicking plasma conditions), with both membrane sides accessible to hydrogenation.

RESULTS AND DISCUSSIONS

In Figure 1 we present a typical snapshot from the MD simulations at an intermediate stage of hydrogenation of h-BN membranes.

In contrast to the case of graphene hydrogenation [5-6], where the graphene membrane retains most of its planar configuration, in the case of h-BN the process is quite distinct. As the hydrogen atoms start to bond to the B and N sites, this process induces significant out-of-plane distortions and the membrane rapidly loses its planar characteristics (see Figure 1). This is a direct consequence of the pronounced sp^3 character created by the hydrogen bonding.

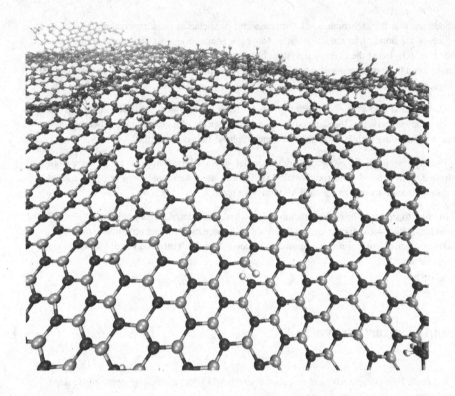

Figure 1. Typical snapshot from MD simulations showing an intermediate stage of the hydrogenation of a h-BN membrane.

The dynamical aspects of the hydrogenation can be better understood analyzing the number of H-B and H-N bonds formed as a function of the time of simulation. These results are presented in Figure 2. The thickness of the curves is a consequence of the fact that the bonds can be formed and dissociated at the same time, which results in fluctuations (recombinations) of their number in time.

As we can see from the Figure, although the number of H-B bonds is slightly larger than the H-N ones, their qualitative behavior in time is similar. At the beginning the rate of hydrogen bonded to the membrane is high, then it decreases and tend to a saturation plateau (more pronounced in the H-N case). In this regime a significant number of defects (broken bonds, vacancies and the removal of N and B atoms) is created and as the process continues it leads to the membrane structural collapse.

These distortions can be partially quantified analyzing the histogram of the second-nearest neighbor distance values. These results are presented in Figure 3. As we can se from the Figure, there is an increase in these distances with relation to the pristine BN values.

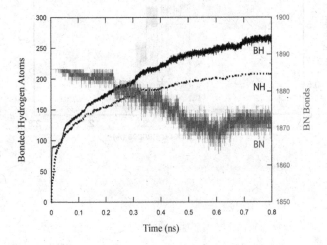

Figure 2. Number of H-B and N-H bonds as a function of the time of simulation. The number of B-N bonds that are not hydrogenated in the system is also indicated.

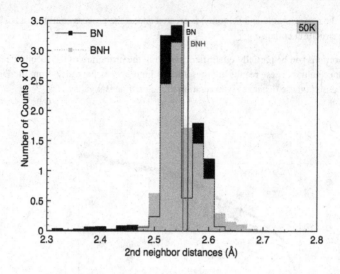

Figure 3. Histogram of the second-nearest neighbor distance values. Mean values are indicated by the continuous lines.

This is a direct consequence of the changes in the bond character, from sp^2 to sp^3. However, due to the significant out-of-plane distortions mentioned above, the net result is that the total membrane dimensions decrease (membrane shrinking). This same shrinking behavior was also reported to hydrogenated [6] and fluorinated [8] graphene membranes. Another important result is that hydrogenated h-BN membranes do not exhibit the formation of correlated domains (islands of hydrogenated atoms), as was observed in the case of graphenes [6].

SUMMARY AND CONCLUSIONS

We have investigated through fully atomistic reactive molecular dynamics simulations the dynamical processes of the hydrogenation of hexagonal boron nitride (h-BN) single-layer membranes.

Our results show that the behavior of H-N and H-B bond formation is similar, but the number of H-B bonds is slightly larger. The out-of-plane deformations caused by the hydrogenation are significantly more pronounced than in the case of graphene membranes under similar conditions. After a critical limit of hydrogen bonds, it was observed the formation of extensive defective areas that can cause the membrane structural collapse. Another important difference with relation to graphane (hydrogenated graphene) is that the existence of correlated domains (islands of hydrogenated atoms) was not observed in the case of h-BN membranes.

ACKNOWLEDGEMENTS

The authors would like to acknowledge financial support from the Brazilian funding agencies Fapesp, CNPq and Fundunesp. The authors would like also to thank Prof. Adri van Duin for many helpful discussions regarding the ReaxFF methodology.

REFERENCES

1. K. S. Novoselov *et al.*, Science **306**, 666 (2004).

2. S. H. Cheng *et al.*, Phys. Rev. B **81**, 205435 (2010).

3. F. Withers, M. Duboist, and A. K. Savchenko, *Phys. Rev. B* **82**, 073403 (2010).

4. J. O. Sofo, A. S. Chaudari, and G. D. Barber, *Phys. Rev. B* **75**, 153401 (2007).

5. D. C. Elias *et al.*, *Science* **323**, 610 (2009).

6. M. Z. S. Flores, P. A. S. Autreto, S. B. Legoas, and D. S. Galvao, *Nanotechnology* **20**, 465704 (2009).

7. R. R. Nair *et al.*, *Small* **6**, 2773 (2010).

8. R. Paupitz, P. A. S. Autreto, S. B. Legoas, S. G. Srinivasan, A. C. T. van Duin, and D. S. Galvao, *Nanotechnology* **24**, 035706 (2013).

9. C. Jin, F. Lin, K. Suenaga, and S. Iijima, *Phys. Rev. Lett.* **102**, 19 (2009).

10. J. C. Meyer, A. Chuvulin, G. Algara-Siller, J. Biskupek, and U. Kaiser, *Nano Lett.* **9**, 2683 (2009).

11. L. Song *et al.*, *Nano Lett.* **10**, 5049 (2010).

12. A. C. T. van Duin, S. Dasgupta, F. Lorant, and W. A. Goddard III, *J. Phys. Chem. A* **105**, 9396 (2001).

13. A. C. T. van Duin and J. S. S. Damste, *Org. Geochem.* **34**, 515 (2003).

14. S. S. Han, J. K. Kang, H. M. Lee, A. C. T. van Duin, and W. A. Goddard III, J. Chem. Phys. **123**, 114703 (2005).

15. S. Plimpton, *J. Comp. Phys.* **117**, 1 (1995). http://lammps.sandia.gov/.

16. R. P. B. dos Santos, E. Perim, P. A. S. Autreto, G. Brunetto, and D. S. Galvao, *Nanotechnology* **23**, 465702 (2012).

Mater. Res. Soc. Symp. Proc. Vol. 1549 © 2013 Materials Research Society
DOI: 10.1557/opl.2013.1055

Mechanical Properties and Fracture Dynamics of Silicene Membranes

Tiago Botari[1], Eric Perim[1], P. A. S. Autreto[1], Ricardo Paupitz[2], and Douglas S. Galvao[1]

[1]Instituto de Física 'Gleb Wataghin', Universidade Estadual de Campinas, 13083-970, Campinas, São Paulo, Brazil.

[2]Departamento de Física, IGCE, Universidade Estadual Paulista, UNESP, 130506-900, Rio Claro, SP, Brazil.

ABSTRACT

The advent of graphene created a new era in materials science. Graphene is a two-dimensional planar honeycomb array of carbon atoms in sp^2-hybridized states. A natural question is whether other elements of the IV-group of the periodic table (such as silicon and germanium), could also form graphene-like structures. Structurally, the silicon equivalent to graphene is called silicene. Silicene was theoretically predicted in 1994 and recently experimentally realized by different groups. Similarly to graphene, silicene exhibits electronic and mechanical properties that can be exploited to nanoelectronics applications.

In this work we have investigated, through fully atomistic molecular dynamics (MD) simulations, the mechanical properties of single-layer silicene under mechanical strain. These simulations were carried out using a reactive force field (ReaxFF), as implemented in the LAMMPS code. We have calculated the elastic properties and the fracture patterns.

Our results show that the dynamics of the whole fracturing processes of silicene present some similarities with that of graphene as well as some unique features.

INTRODUCTION

The chemistry of carbon is very rich, the three different hybridization states (sp, sp^2 and sp^3) allow the generation of a large number of different structures, such as: diamond (sp^3), graphite, graphene (single-layer graphite), fullerenes and nanotubes (sp^2) [1], and graphynes (sp) [2-4].

The carbon-based structures of reduced dimensionality (such as, fullerenes and nanotubes) have been shown to exhibit some extraordinary structural, thermal and electronic properties. Another spectacular example of this is graphene [5]. The advent of graphene created a new era in materials science. Due to its unique electronic and mechanical properties, graphene is considered as the basis for a new electronics [5-7]. However, in its pristine form graphene is a zero bandgap semiconductor, what limits its use in some transistor applications [7]. In part because of this, there is a renewed interest in other possible graphene-like structures, based on carbon or other chemical elements.

Graphene (see Figure 1) is a two-dimensional (planar) honeycomb array of carbon atoms. A natural question is whether other IV-group elements of the periodic table (such as, silicon and germanium), could also form graphene-like structures. For these elements the graphene equivalent structures are called silicene [8] (see Figure 1) and germanene [9].

Silicene, a single-layer two-dimensional honeycomb silicon sheet, was first predicted to exist based on ab initio calculations in 1994 [8]. It was recently synthesized by different groups [10-15]. Germanene has been also experimentally realized [16].

Silicene shares some of the attractive electronic properties exhibited by graphene, making it a very promising material for electronics and spintronics applications. Another advantage would be that all the silicon present-day technology, that is the basis of our electronics [17], could be easily adapted to silicene integration. One peculiar aspect of silicene it that it is not a true two-dimensional material, as a significant level of buckling is always present [8-16].

In this work we have investigated, using classical molecular dynamics simulations, the mechanical properties and fracture (mechanical failure) dynamics of single-layer silicene membranes under mechanical strain.

Our results show that the dynamics of fracture of silicene membranes present some similarities with that of graphene as well as some unique features.

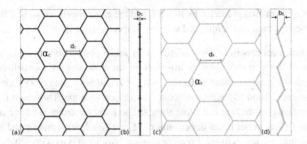

Figure 1. Schematic view of graphene and silicene membranes in the same scale. (a/c) and (b/d) refer to frontal and lateral view of graphene and silicene, respectively.

METHODOLOGY

We have carried out fully atomistic molecular dynamics simulations in order to investigate mechanical properties and fracture dynamics of silicene under strain load. We have used the reactive force field ReaxFF [18], as implemented in the large-scale atomic/molecular massively parallel simulator (LAMMPS) code [19]. We have used NPT and NVT ensembles and the temperature of the simulations was controlled using a Nosé-Hoover thermostat, as implemented in the LAMMPS code [19]. The simulations were carried out at 10K and 150 K. These temperatures were chosen because they are more appropriate to study the buckling/unbuckling mechanisms. For higher temperatures the thermal fluctuations make this analysis more difficult. We have used time-steps of 0.05 fs and the simulations were run until complete mechanical failure (breaking) of the membranes.

This methodology was successfully used to investigate the mechanical failure of carbon-based nanostructures such as graphene and carbon nanotubes [20].

ReaxFF [18] is a reactive force field developed by van Duin, Goddard III and co-workers and can be used in MD simulations of large systems. It allows simulations of many types of chemical reactions. It is similar to standard non-reactive force field, like MM3 [18], where the system energy is divided into partial energy contributions associated with, amongst others, valence angle bending, bond stretching, and non-bonded van der Waals and Coulombic interactions. However, one main difference is that ReaxFF can handle bond formation and dissociation (making/breaking bonds) as a function of the bond order values. ReaxFF was parameterized against density functional theory (DFT) calculations, being the average deviations between the heats of formation predicted by ReaxFF and the experiments, equal to 2.8 and 2.9 kcal/mol, for non-conjugated and conjugated systems, respectively [18].

Our silicene structures consist of semi-infinite strips, under periodic boundary conditions. Firstly, in order to obtain a relaxed (zero strain) structure, the structures are thermalized using a NPT ensemble with the pressure value set to zero. After that and now using a NVT ensemble at different temperatures (10 and 150 K) the structures are subject to strain. This strain is generated by the gradual increasing of the unit cell value along the periodic direction. We have considered silicene strips with typical initial linear dimensions of 95 and 100 Å for armchair and zigzag edge terminated structures, respectively. From topological reasons, it is expected that the fracture dynamics will be different depending on edge termination and applied strain direction. This will be addressed in more details in the results and discussions section. A constant strain rate of 2.0×10^{-6} s^{-1} was continuously applied until the mechanical rupture.

One effective way to investigate the dynamics of the mechanical failure is through the analysis of how the stress is accumulated and dissipated when the structures are subjected to strain. In order to carry out this analysis we calculated the stress response to the applied strain. The stress is calculated by computing the forces on each atom, then obtaining the stress tensor. From this tensor we calculated a quantity known as *von Mises* stress [21], which is the second invariant of the deviatoric stress tensor. It is related to distortion state of the system and provides very helpful information on the fracturing processes because it is possible to visualize where the stress is accumulating and being dissipated, both in space and time.

RESULTS AND DISCUSSIONS

In Figure 1 we present, in the same scale, the ReaxFF optimized geometries for graphene and silicene structures. For silicene the obtained Si-Si bond distance values were of $d_{Si-Si} = 2.3$ Å, buckling (out of plane) value of $b_{Si} = 0.6$ Å and bond angle value (see Figure 1) of $\alpha_{Si} = 111°$. These values are in good agreement with DFT results reported in the literature [8, 9]: $d_{Si-Si} = 2.5$ Å, $b_{Si} = 0.44$ Å. The ReaxFF results for graphene are: $d_{C-C} = 1.4$ Å, $b_{Si} = 0.0$ Å, $\alpha_{Si} = 120°$.

Once obtained the optimized geometries, we then proceeded to the stress/strain calculations for different temperatures. In Figure 2 we present the results for zigzag and armchair edge terminated structures for the temperature of 150 K. As we can see from the Figure, for both, zigzag and armchair structures, the stress increases almost linearly with the applied strain until a rupture (mechanical failure) occurs. The main differences are that these ruptures occur about 15% and 30% of applied strain for armchair and zigzag structures, respectively.

The Young's modulus was estimated using a linear regression fitting to the linear part of the stress-strain curves (see Figure 2). The obtained values for the armchair and zigzag structures were both of 0.042 TPa nm. This value is ~7.6 times smaller than the corresponding value to graphene under similar conditions [22, 23].

In Table 1 we present the critical strain values ε_c (where the mechanical failure occurs) for the different structures and for the temperatures considered in this study. As we can see from the table, the threshold strain for the plastic regime is highly dependent on the temperature values.

The significant difference in the critical strain value can be explained as a consequence of the hexagonal rings structures and the directions of the applied strain. For the armchair structures the forces are applied along the direction of some of the chemical bonds composing the structures, while for the zigzag ones this does not happen, which causes the mechanical deformations to produce an increase/decrease of the angle values on the rings. This causes the forces to be redistributed, in arch-type-like effect, thus making the structures more resilient to mechanical deformation.

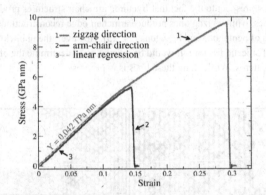

Figure 2. Stress versus strain curves for zigzag and armchair edge terminated structures. Results for the temperature of 150 K. See text for discussions.

Temperature (K)	ε_c (zigzag) (GPa nm)	ε_c (arm-chair) (GPa nm)
10	0.35	0.20
150	0.30	0.15

Table 1. Critical strain ε_c as a function of temperature, for zigzag and armchair structures.

As mentioned above the von Mises stress distributions for the stretched structures provide valuable information on the dynamics of the fracture mechanisms. In Figures 3 and 4 we present these results for the armchair and zigzag structures at different states of the rupture processes. We can see that the stress distribution is much more uniform in the case of zigzag nanoribbons than in the case of armchair nanoribbons. This could explain why armchair nanoribbons present a higher level of edge reconstruction.

In Figure 3 we present typical results for a zigzag structure. As we can see from that figure, the rupture creates a clean and well-formed armchair edge terminated structures. Only very few pentagon and heptagon reconstructed rings were observed.

In Figure 4 we present the corresponding results for the armchair structures. As we can see from that figure, in this case the fractured structure present clear and well-formed zigzag edge (in contrast with the armchair ones of the previous case) terminated structures. Also, the number of reconstructed pentagon and hexagon rings is more significant than in the case of zigzag structures.

These fracture patterns share some similarities with those observed for graphene structures [22,23], especially with respect to the fact that fractured armchair structures produce edge terminated zigzag ones, while zigzag ones produce armchair edge reconstructions. Both case exhibit the formation of pentagons and hexagons [24, 25]. However, the unbuckling mechanism is a unique feature of silicene fracture. It would be interesting to determine the effect of different substrates on these patterns. Work along these lines is in progress.

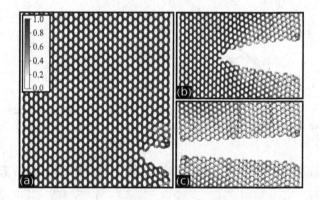

Figure 3. Typical snapshots from MD simulations showing different stages of the mechanical failure of a zigzag silicene membrane under mechanical strain. Darker regions indicate high stress values.

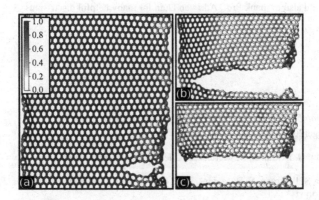

Figure 4. Typical snapshots from MD simulations showing different stages of the mechanical failure of an armchair silicene membrane under mechanical strain. Darker regions indicate high stress values.

SUMMARY AND CONCLUSIONS

In summary, we have investigated, using fully atomistic reactive (ReaxFF force field) molecular dynamics simulations, the structural and mechanical properties of single-layer silicene (the silicon structure equivalent to graphene) membranes under mechanical strain. The estimated Young's modulus was of Y=0.042 TPa nm for both armchair and zigzag edge terminated structures. The fracture patterns share some features with the ones reported to graphene membranes (fractured armchair structures produce edge terminated zigzag ones and vice-versa), but present unique ones, such as the unbuckling mechanisms.

ACKNOWLEDGEMENTS

The authors acknowledge financial support from the Brazilian Agencies FAPESP, CNPq and CAPES. We would also to thank Prof. Adri van Duin for many helpful discussions.

REFERENCES

1. *Carbon Nanotube Science*, Peter J. F. Harris, Cambridge University Press, Cambridge(2009).

2. R. Baughman, H. Eckhardt, M. Kertesz, *J. Chem. Phys.* **87**, 6687 (1987).

3. V. R. Coluci, S. F. Braga, S. B. Legoas, D. S. Galvao, and R. H. Baughman, *Phys. Rev. B* **68**, 035430 (2003).

4. V. R. Coluci, S. F. Braga, S. B. Legoas, D. S. Galvao, and R. H. Baughman, *Nanotechnology* **15**, S142 (2004).

5. K. S. Novoselov *et al.*, *Science* **306**, 666 (2004).

6. S. H. Cheng *et al.*, *Phys. Rev. B* **81**, 205435 (2010).

7. F. Withers, M. Duboist, and A. K. Savchenko, *Phys. Rev. B* **82**, 073403 (2010).

8. K. Takeda and K. Shiraishi, *Phys. Rev. B* **50**, 14916 (1994).

9. S. Cahangirov, M. Topsakal, E. Akturk, H. Sahin, and S. Ciraci, *Phys. Rev. Lett.* **102**, 236804 (2009).

10. H. Nakano *et al.*, *Angew. Chem.* **118**, 6451 (2006).

11. B. Lalmi *et al.*, *Appl. Phys. Lett.* **97**, 223109 (2010).

12. G. M. Psofogiannakis and G. E. Froudakis, *J. Phys. Chem. C* **116**, 19211 (2012).

13. B. Aufray *et al.*, *Appl. Phys. Lett.* **96**, 183101 (2010).

14. P. de Padova *et al.*, *Appl. Phys. Lett.* **96**, 261905 (2010).

15. P. Vogt *et al.*, *Phys. Rev. Lett.* **108**, 155201 (2012).

16. E. Bianco *et al.*, *Nano Lett.* **7**, 4414 (2013).

17. R. Friedlein, A. Fleurence, T. Ozaki, and Y. Yamada-Takamura, *SPIE Newsroom*, *in press* DOI : 10.1117/2.1201305.004854.

18. A. C. T. van Duin, S. Dasgupta, F. Lorant, and W. A. Goddard III, *J. Phys. Chem. A* **105**, 9396 (2001).

19. S. Plimpton, *J. Comp. Phys.* **117**, 1 (1995). http://lammps.sandia.gov/.

20. R. Paupitz *et al.*, *Nanotechnology* **24**, 035706 (2013).

21. Y. Yang and X. Xu, *Comp. Mater. Sci.* **61**, 83 (2012).

22. Q. X. Pei, Y. W. Zhang, and V. B. Shenoy, *Carbon* **48**, 898 (2010).

23. K. Kim *et al.*, *Nano Lett.* **12**, 293 (2011).

24. Koskinen et al., Phys. Rev. Lett. **101**, 115502 (2008).

25. Koskinen et al., Phys. Rev. B 80, 073401 (2009).

Carbon Nanotubes

Mater. Res. Soc. Symp. Proc. Vol. 1549 © 2013 Materials Research Society
DOI: 10.1557/opl.2013.965

Covalent Chemical Modification of Single-walled Carbon Nanotubes Using Azide Functionalised Anthraquinone Derivatives for Pseudocapacitor Application.

Charlotte Frayère, Christophe Galindo, Laurent Divay, Michel Paté, Pierre Le Barny.
Thales Research and Technology-France, Campus Polytechnique, 1 Avenue Augustin Fresnel, 91 767 Palaiseau , France.

ABSTRACT

Electrodes made of single-walled carbon nanotubes (SWCNTs) chemically modified by a series of anthraquinone derivatives (AQ) have been prepared and characterized by cyclic voltammetry in 0.1M H_2SO_4, using the standard 3 electrode set-up and by Raman spectroscopy. It has been demonstrated that a AQ modified SWCNT electrode provided between 114 to 220% higher specific capacitance, compared to pristine SWCNT electrode, depending on the length of the spacer between SWCNT and AQ.

INTRODUCTION

It is well known that the world fossil fuels reserves are limited. This situation dictates that alternative renewable energy resources be developed and implemented. However, renewable energy is inherently intermittent and fluctuating. One of the best ways to develop efficient energy storage technologies, is to convert chemical energy into electrical energy. Today, fuel cells, batteries and supercapacitors are the main electrical energy storage systems which are commercially available. Each storage system has a suitable application range. Thus, supercapacitors also known as Electrochemical Double Layer Capacitors (EDLCs), fill the gap between conventional batteries and capacitors. They exhibit higher power densities but lower energy densities than batteries while exhibiting higher energy densities but lower power densities than capacitors. Activated carbon (AC) exhibiting very high surface area (1000-2000m²/g) are currently used as electrode material in commercially available supercapacitors. Electrolyte can be either aqueous or organic [1,2]. To increase the specific capacitance of a carbon electrode, an electrochemically active material can be deposited or covalently grafted onto the surface of the active material, thus defining what is called the pseudo-capacitive behavior. Two types of electroactive materials are currently under study : metal oxides such as RuO_2 [3] or MnO_2 [4,5] and organic compounds such as conducting polymers [6] or electroactive molecules. Among these, chemical modification of carbon fabric [7] or activated carbon [8-10] with anthraquinone (AQ) derivatives has been the subject of investigations these last few years. In all cases, carbonaceous material was functionalized via diazonium salts, leading to carbon – carbon bonds between AQ and the electrode. It has been demonstrated that although the chemical modification of activated carbon decreased its specific surface area because of the grafting which occurred primarily at the entrance of the ultramicropores and micropores, the specific capacitance of the electrode increased significantly compared to the pristine electrode [10].

Even though activated carbon is currently the state of the art electrode material, it suffers from some drawbacks including its limited electrical conductivity and accessibility of its pores to solvated ions of the electrolyte and the need for a binder to ensure the expected electrode cohesion. These drawbacks can be overcome when carbon nanotubes (CNTs) are used instead of activated carbon. Indeed, CNTs have a rather low resistivity in the range of 5-50μΩcm [11] and a quite high surface area (e.g. 300-400m²/g). Finally, CNTs can be easily transformed into a

111

nanotube network also called bucky paper without the use of a binder [12]. The pore size distribution of the bucky paper is centred around 10nm which is convenient when compared to the size of the solvated ions, namely around 1nm [13,14].

In the present study we report the preparation and the characterization of electrodes in the form of bucky papers, made of single-walled carbon nanotubes (SWCNTs) chemically modified by a series of azido functionalised anthraquinone derivatives (Figure 1). Thermal decomposition of the azido derivatives leads to nitrenes which give a [2+1] cycloaddition reaction with CNTs [15].

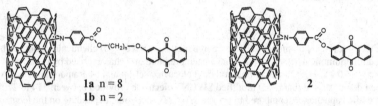

1a n = 8
1b n = 2

2

Figure 1 : Structure of the chemically modified SWCNTs under study.

Distance between the AQ moiety and the SWCNTs has been varied, thus allowing the study of the influence of the distance between the electroactive molecule and the SWCNTs on the electrochemical properties of the resulting electrode.

EXPERIMENT

AQ modified SWCNTs **1a**, **1b**, **2** were prepared from azido derivatives **3**, **4** and **5** respectively, according to the adapted synthetic route described in [16]. The overall pathway of compounds **3**, **4**, and **7** are given in Figure 2. SWCNTs were purchased from Bucky USA (BU203, purity 90% in weight, length 0.5-5 μm, diameter 0.7-2.5 nm). ^1H and ^{13}C NMR were measured on a Bruker Avance III 400 MHz spectrometer by using tetramethylsilane (TMS; $\delta =$ 0ppm) as an internal standard. FTIR spectra were recorded on a Thermo-Nicolet Nexus spectrometer.

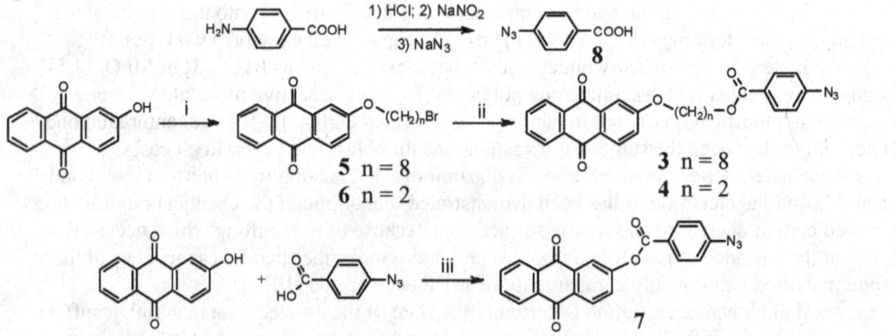

Figure 2 : Synthesis of AQ derivatives functionalised with an azido group.
(i. n=8 KOH, EtOH; n = 2 K_2CO_3, cyclohexanone. ii. DBU, DMPU. iii. DCCI, DPTS, THF)

The melting points were measured with a DSC-7 apparatus from Perkin-Elmer at a heating rate of $10°C.min^{-1}$. Thermogravimetric characterizations were performed with a TGA-7 apparatus from Perkin-Elmer under nitrogen atmosphere at a heating rate of $10°C.min^{-1}$.

AQ modified SWCNTs were characterized by cyclic voltammetry (BioLogic VMP3 potentiostat) in $0.1M$ H_2SO_4, using the standard 3 electrode set-up at a scan rate of $10mVs^{-1}$. The working electrode was a few mg in weight of bucky paper placed in a stainless steel grid used as a current collector. Platinum was chosen as the counter electrode and SCE as reference electrode.

Raman spectra were measured with a Renishaw Raman spectrometer using 514.5nm line of an argon ion laser with power of $200\mu W$.

Synthesis of 4-azidobenzoic acid 8

4-azidobenzoic acid was prepared according to the adapted synthetic route described in [17]. Yield 1.10g (67.5%).

IR (ATR) : $v = 2108cm^{-1}$ ($-N_3$), 1679 cm^{-1} ($-C=O$ acid), 1602 cm^{-1} ($C=C$). ^1H NMR (DMSOd$_6$) δ 7.2-8.0 (*para* AA'BB' q, J = 8.8Hz aromatic).

Synthesis of 2-(8-bromooctyloxy)-anthraquinone 5

2g (8.92 mmol) of 2-hydroxyanthaquinone were dissolved in a solution of 0.5g of KOH in 20mL of ethanol. Then 6.5g 22.64 mmol) of 1,8-dibromooctane were added and the reaction mixture was heated under reflux for 14h. After removal of the solvent, the crude product was purified by column chromatography on silica gel by using hexane and then hexane/methylene chloride (1:1, v/v) as the eluent to afford pure **5** as a pale yellow solid (1,71g, 46%). Mp : 80°C. IR (ATR) : $v = 1669cm^{-1}$ ($C=O$), $1589.9cm^{-1}$ ($C=C$). ^1H NMR (CDCl$_3$) δ 1.36-1.43 (8H, m, CH_2), 1.84 (4H, m, CH_2), 3.40 (2H, t, CH_2), 4.10 (2H, t, CH_2), 7.18-7.21 (1H, m, aromatic), 7.64-7.76 (3H, m, aromatic), 8.24-8.27 (3H, m, aromatic), ^{13}C NMR (CDCl$_3$) δ 25.9, 28.1, 28.7, 29.1, 32.8, 33.8, 68.8, 110.5, 121.4, 126.9, 127.1, 129.7, 133.6, 133.7, 134.1, 135.6, 163.9, 182.0, 183.2.

Synthesis of 2-(2-bromoethyloxy)-anthraquinone 6

6 was prepared according to the procedure described for **5**, using K_2CO_3 as base and cyclohexanone as solvent.Yield : 59.3%. Mp : 138°C.

IR (ATR) : $v = 1675cm^{-1}$ ($C=O$), 1590 cm^{-1} ($C=C$). ^1H NMR (CDCl$_3$) δ 3.27 (2H, t, CH_2 J = 6 H$_z$), 4.50 (2H, t, CH_2 J = 6.1 Hz), 7.28-7.32 (1H, m, aromatic), 7.71-7.82 (3H, m, aromatic), 8.26-8.29 (3H, m, aromatic), ^{13}C NMR (CDCl$_3$) δ 28.4,. 68.3, 110.5, 121.6, 127.2, 127.7, 129.9, 133.5, 133.6, 133.7, 134.2, 135.6, 162.8, 182.0, 183.0.

Synthesis of 2-(4-azidobenzoyloxyoctyloxy)-anthraquinone 3

3 was prepared according to the adapted synthetic route described in [18]. The crude product was purified by column chromatography by using hexane/methylene chloride (1:1, v/v) as the eluent to afford a pale yellow solid (0,82g, 91.6%). Mp : 99°C. Decomposition onset 170°C

IR (ATR) : $v = 2132cm^{-1}$ ($-N_3$), 1707 cm^{-1} ($-C=O$ ester), $1672cm^{-1}$ ($C=O$), 1602 cm^{-1} ($C=C$) $1591.cm^{-1}$ ($C=C$ AQ). ^1H NMR (CDCl$_3$) δ 1.3-1.6 (8H, m, CH_2), 1.7-1.9 (4H, m, CH_2), 4.14 (2H, t, CH_2), 4.32 (2H, t, CH_2), 7.04 (2H,d, aromatic), 7.22-7.25 (1H, m, aromatic), 7.69-7.78 (3H, m, aromatic), 8.02 (2H, d, aromatic), 8.2-8.35 (3H, m, aromatic), ^{13}C NMR (CDCl$_3$) δ 25.9, 25.97, 28.1, 28.7, 29.01, 29.17, 29.2, 65.1.8, 68.8, 110.6, 118.83, 121.5, 127.1, 129.7, 131.4, 133.6, 133.7, 133.8, 134.1, 135.6, 144.69, 164.0, 165.8, 182.1, 183.3.

Synthesis of 2-(4-azidobenzoyloxyethyloxy)-anthraquinone 4

By a similar procedure to that described for the synthesis of **3**, **4** was obtained in 77.3% yield. Mp : 152.5°C. Decomposition onset 172°C. IR (ATR) : $v = 2143cm^{-1}$ ($-N_3$), 1712 cm^{-1} ($-C=O$ ester), $1678cm^{-1}$ ($C=O$), 1598 cm^{-1} ($C=C$). ^1H NMR (CDCl$_3$) δ 4.51 (2H, t, CH_2), 4.73 (2H, t,

CH_2), 7.04 (2H,d, aromatic), 7.30-7.33 (1H, m, aromatic), 7.77-7.80 (3H, m, aromatic), 8.05 (2H, d, aromatic), 8.2-8.35 (3H, m, aromatic), ^{13}C NMR (CDCl$_3$) δ 63.0, 66.6, 110.6, 118.9, 121.7, 126.2, 127.2, 127.5, 129.9, 131.6, 133.6, 133.7, 134.2, 135.6, 145.1, 163.3, 165.6, 182.1, 183.1

Synthesis of 2-(4-azidobenzoyloxy)-anthraquinone 7

3 was prepared according to the adapted synthetic route described in [19]. The crude product was purified by column chromatography by using hot CHCl$_3$ as the eluent to afford a pale yellow solid. Yield 78.3%. Mp : 196°C decomposition.

IR (ATR) : v = 2135cm^{-1} (-N$_3$), 1739 and 1730 cm^{-1} (-C=O ester), 1675cm^{-1} (C=O), 1602 cm^{-1} (C=C), 1590.cm^{-1} (C=C AQ). ^{1}H NMR (CDCl$_3$) 7.16 (2H,d, aromatic), 7.65-7.68 (1H, m, aromatic), 7.8-7.83 (2H, m, aromatic), 8.13 (1H, d, aromatic), 8.21 (2H, m, aromatic), 8.32-8.34 (2H, m, aromatic), 8.41 (1H, d, aromatic), ^{13}C NMR (CDCl$_3$) 119.2, 120.2, 125.1, 127.4, 127.5, 129.4, 131.2, 132.2, 133.46, 133.53, 134.2, 134.3, 135.3, 146.1, 155.5, 163.6, 182.2, 182.3.

General procedure of surface modification of SWCNTs [16]

SWCNTs (50mg) were dispersed in o-dichlorobenzene (50mL) using an ultrasonic bath for 3 h. To this suspension was added 0.56 mmol of azide functionalised AQ under argon and the reaction mixture was heated at 135°C for 24 h. After cooling to RT, the functionalised SWCNTs were collected by centrifugation at 3000 rpm and extensively washed with THF, CHCl$_3$ and Et$_2$O.

Bucky paper preparation

A mixture of 28mg of modified SWCNT in 90mL of NMP was sonicated for 3hours. The obtained suspension was centrifuged at 3000rpm for 20 minutes. The supernatant was then filtered off under vacuum through a 0.2μm Anodis 25 filter (Sartorius Stedim). The buckypaper thus formed was extensively washed with CH$_2$Cl$_2$, acetone and Et$_2$O and finally dried under vacuum overnight.

Electrochemical characterization

Cyclic voltamograms (CV) in 0.1M H$_2$SO$_4$ of electrodes made with bucky paper prepared from **1a**, **1b**, and **2** are characterized by a rectangular shaped due to capacitive behaviour of the SWCNTs surface, completed by reversible redox waves for the AQ (Figure 3). We have estimated the amount of grafted AQ by integrating the anodic or the cathodic redox waves, taking into account that the redox process of AQ involves 2 protons and 2 electrons. The results are summarised in Table 1.

Compound	E_{ox} (mV)	E_{red} (mV)	Cs Fg^{-1}	AQ grafted wt %	AQ grafted μmole/g of C
1a	-4	-390	75	11.3	271
1b	-19	-369	105	13.6	409
2	54	-384	112	14.2	486

Table 1 : Properties of electrodes under study. Maximum oxidation potential (E_{ox}), maximum reduction potential (E_{red}), specific capacitance (Cs), AQ grafting (wt %) and AQ grafting (ratio of the amount (μmol) of AQ to the mass of SWCNT on the electrode.

A shift towards higher potential was observed for the maximum oxidation potential (E_{ox}) and the maximum reduction potential (E_{red}) of **2** compared to the behaviour of **1a** and **1b**. This was attributed to the fact that AQ moiety was linked to an aromatic ester, a weak electron withdrawing group, instead of an ether which is an electron donor group.

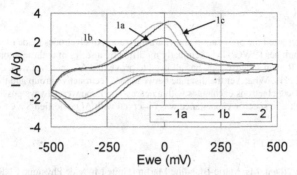

Figure 3 : Cyclic voltamogram of **1a,1b** and **2** recorded in 0.1M H_2SO_4 (scan rate : 10mVs⁻¹).

DISCUSSION

The covalent functionalization of SWCNTs was confirmed by Raman spectroscopy as shown in Figure 4 where the typical peaks at 1589 cm⁻¹ and 1352 cm⁻¹ attributed to G and D band , respectively. The ratio D band intensity over G band intensity I_D/I_G increased for all functionalised samples, with comparison of pristine SWCNTs. This is the signature of the presence of scattering defects on sidewall of SWCNTs due to covalent functionalization.

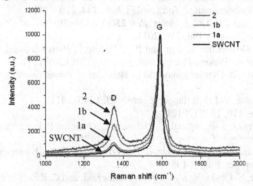

Figure 4 : Raman spectra of pristine and functionalized SWCNTs with AQ derivatives.

As expected, the ratio I_D/I_G increase was pretty well related to the increase of the grafting rate expressed in μmol/g of carbon. It appears that even if the electroactive moiety is far from the carbon nanotube on which it is grafted (**1a**), it can exchange electrons with a nearby nanotube.

Compared to pristine SWCNT whose specific capacitance is 35F/g, AQ modified SWCNT electrode provided between 114 to 220% higher specific capacitance depending on the length of the spacer.

CONCLUSIONS

We have reported the preparation and characterization of electrodes made of single-walled carbon nanotubes (SWCNTs) chemically modified by a series of anthraquinone derivatives (AQ). Raman spectroscopy has allowed to prove that AQ moieties were covalently linked to the SWCNTs. Whatever the distance between the electroactive group and the SWCNT on which it is grafted, electrons exchanges with a nearby nanotube were always possible. Finally, a significant increase in specific capacitance was observed with AQ modified SWCNTs electrodes

ACKNOWLEDGMENTS

The authors thank Ms. Marie-Blandine Martin (Unité Mixte de Physique CNRS-Thales UMR137) and Dr. Bernard Servet (Thales Research and Technology) for Raman analysis.

REFERENCES

1. P. Simon, and Y. Gogotsi, *Nat. Mater.* **7**, 845 (2008).
2. J. W. Long, D.l Bélanger, T. Brousse, W. Sugimoto, M. B. Sassin, and O. Crosnier, *MRS Bulletin* **36**, 513 (2011)
3. B. E. Conway, *Electrochemical Supercapacitors: Scientific Fundamentals and Technological Applications*, (Kluwer Academic/Plenum Publishers , New York, 1999).
4. H. Y. Lee, and J. B. Goodenough, *J. Solid State Chem.* **144**, 220 (1999).
5. C. Xu, F. Kang, B. Li, and H. Du, *J. Mater. Res.* **25**(8), 1421 (2010)
6. A. Laforgue, *J. Power Sources* **196**, 559 (2011)
7. K. Kalinathan, D. P. DesRoches, X. Liu, and P. G. Pickup, *J. Power Sources* **181**, 182 (2008)
8. R. D. L. Smith, and P. G. Pickup, *Electrochim. Acta* **54**, 2305 (2009)
9. G. Pognon, T. Brousse, L. Demarconnay and D. Bélanger, *J. Power Sources*, **196**, 4117 (2011)
10. G. Pognon, T. Brousse, and D. Bélanger, *Carbon* **49**, 1340 (2011)
11. D. Eder, *Chem. Rev.* **110**, 1348 (2010)
12. P. Bondavalli, D. Pribat, J.-P. Schnell, C. Delfaure, L. Gorintin, P. Legagneux, L. Baraton, and C. Galindo, *Eur. Phys. J. Appl. Phys.* **60,** 10401-p1 (2012)
13. C. Vix-Guterl, S. Saadallah, K. Jurewicz, E. Frackowiak, M. Reda, J. Parmentier, J. Patarin, and F. Béguin, Mater. Sci. Eng. B **108**, 148 (2004)
14. Y. J. Kim, Y. Horie, S. Ozaki, Y. Matsuzawa, H. Suezaki, and C. Kim, *Carbon* **42**, 1491 (2004)
15. D. Tasis, N. Tagmatarchis, A. Bianco, and M. Prato, *Chem. Rev.* **106**, 1105 (2005)
16. N. Mackiewicz, J. A. Delaire, A. W. Rutherford, E. Doris, and C. Mioskowski, *Chem. Eur. J.* **15**, 3882 (2009)
17. M. Minato, and P. M. Lahti, *J. Am. Chem. Soc.* **119**(9), 2187 (1997)
18. N. Ono, T. Yamada, T. Saito, K. Tanaka, and A. Kaji, *Bull. Chem. Soc. Jpn.* **51**(8), 2401 (1978)
19. J. Luo, M. Haller, H. Ma, S. Liu, T.-D. Kim, Y. Tian, B. Chen, S.-H. Jang, L. R. Dalton, and A. K.-Y. Chen, *J. Phys. Chem. B* **108**, 8523 (2004)

Mater. Res. Soc. Symp. Proc. Vol. 1549 © 2013 Materials Research Society
DOI: 10.1557/opl.2013.1032

Determining In-plane and Thru-plane Percolation Thresholds for Carbon Nanotube Thin Films Deposited on Paper Substrates Using Impedance Spectroscopy

Rachel L. Muhlbauer[1] and Rosario A. Gerhardt[1]
[1]Georgia Institute of Technology, 771 Ferst Dr.
Atlanta, GA 30332, USA

ABSTRACT

Concentration- and layer-dependent percolation thresholds can be determined for carbon nanotube (CNT) films deposited from aqueous dispersions on paper substrates at both the surface of the deposited film (in-plane) and through the thickness of the paper (thru-plane) using impedance spectroscopy. By analyzing the impedance spectra as a function of the number of layers (solution concentration is constant) or the solution concentration (number of layers is constant), the electrical properties and percolation thresholds for CNT-paper composites can be determined. In-plane measurements show that percolation occurs at 4 layers when 1 mg/mL solution concentration is used. In the thru-plane direction, the films are already percolated at 1 mg/mL concentration, which is confirmed by varying the concentration of the solution used to deposit 1 layer films. A second percolation event happens between 8 and 12 layers due to an increased number of interconnections of CNTs within the paper substrate. The lowest sheet resistance achieved was 100 $\Omega/\square$.

INTRODUCTION

In recent years, paper electronics have soared in popularity due to the low-cost, widespread availability, and flexible nature of paper [1]. At the same time, research into the use of carbon nanotubes (CNTs) as an electronic conducting material has grown due to the exceptional theoretical properties exhibited by CNTs. Printing as a film deposition method for CNTs onto flexible paper substrates has been shown previously to be a cost effective method of creating electronically conductive films for a variety of applications [2-4]. If the application does not require complex shapes, it is possible to print highly conductive films of non-surface modified multiwalled carbon nanotubes (MWNTs) from an aqueous dispersion on paper substrates using only a few printed layers by using a vacuum filtration-aided dropcasting technique [5,6].

Additionally, it can be shown how the electrical properties evolve both on the surface (in-plane) and through the thickness (thru-plane) of the paper through the use of impedance spectroscopy. Impedance spectroscopy is an AC electrical characterization technique which is used to determine frequency dependent electrical behavior in films. AC electrical techniques are more sensitive to individual electronic processes present in films making it possible to separate contributions of these different electronic processes from the overall electrical behavior of the film [5-7]. While we will not discuss each electrical process found in the films in detail in this paper, we will show how we can use the data found by impedance spectroscopy to determine percolation thresholds for these MWNT-paper films made as described.

EXPERIMENT

Aqueous dispersions of multiwalled carbon nanotubes (MWNT) obtained from cheaptubes.com (0.5-2.0 μm in length, 8-15 nm in diameter, >95 wt% purity with <1.5 wt% ash) and dispersed using sodium dodecylbenzenesulfonate (SDBS) surfactant were used to deposit the films in this study. 0.1, 0.5, and 1 mg/mL MWNT dispersions were stabilized with 10 mg/mL of

SDBS in DI water using the same method discussed in [5]. 5 mg/mL solutions were made with the exact same MWNTs, DI water, and mixing steps except that 100 mg/mL of the SDBS surfactant was required to achieve a stable dispersion of the MWNTs in water. Each film described in this paper was obtained by dropcasting 150 μL of the MWNT dispersion per layer onto the 413 qualitative filter paper from VWR.com (5 μm pore size) while under vacuum filtration [5]. Each layer was allowed to completely dry before the next layer was deposited. Three different film sets were made: 1) 1-20 layers made from 1 mg/mL dispersions and characterized both in-plane and thru-plane; 2) 4 layers made with all the concentrations and characterized in-plane; 3) 1 layer made with all the concentrations and characterized thru-plane.

Characterization of these samples was done by a combination of AC electrical techniques and imaging techniques. For the in-plane AC measurements, two impedance spectroscopy configurations were used. In both setups, a four point probe stage with a probe spacing of 1.5875 mm was used and the high voltage/current and low voltage/current cables were shorted together such that only two probes were used per measurement. For the very insulating samples, a Solartron 1296 connected to a Solartron 1260 was used. For the less resistive samples, only the Solartron 1260 was needed. For the thru-plane measurements, the samples were loaded into a parallel plate arrangement and measured using the high resistance impedance setup. In all cases, frequencies from 20 MHz to 0.1 Hz were measured under an AC voltage of 0.1 V. In-lens SEM images were taken with a Zeiss SEM Ultra60 at 1-2 kV to determine the microstructure of the films in-plane.

RESULTS AND DISCUSSION

Using vacuum filtration as a tool during film deposition leads to a film formation which is very different than inkjet printing. Vacuum filtration is a directional deposition method that pulls the CNT dispersions through the paper thickness, resulting in CNT deposition both on the surface (in-plane) but also within the thickness of the paper substrate (thru-plane). Film formation, therefore, becomes a function of the pore size of the substrate. Pore sizes similar to the length of the CNTs aid in film deposition on the surface as fewer CNTs penetrate into the paper structure. Conversely, when pore sizes are much larger than the length of the CNTs, more of the dispersion is pulled through the paper substrate and less deposits on the surface [5]. In the case of this study, the pores of the paper (5 μm) are slightly larger than the length of the CNTs (0.5-2 μm) allowing for dual deposition to occur: CNTs are interconnected both through the thickness and on the surface of the paper substrate. The extent to which the CNTs are interconnected, however, depends on the number of layers deposited or the concentration of the dispersion used. Figure 1 shows the complex impedance data for all of the experiments done in this study (shown in log-log form in order to show all the graphs per study on one image). The top row shows the impedance results for the in-plane measurements, where 1a shows the data for the different number of layers deposited at constant concentration and 1b shows the data for 4-layered films deposited at different concentrations. The bottom row of figure 1 shows the impedance results for the thru-plane measurements, where 1c shows the data for the different number of layers deposited at constant concentration and 1d shows the data for 1-layered films deposited at different concentrations.

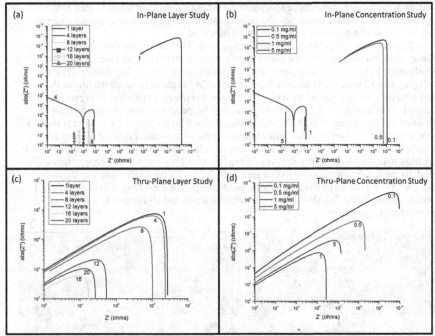

Figure 1: Impedance spectroscopy results (shown on log-log scale) for a) the in-plane layer study; b) the in-plane concentration study; c) the thru-plane layer study; and d) the thru-plane concentration study

In-Plane

Figure 1a shows the impedance spectroscopy results for the in-plane layer study done at a constant concentration of MWNTs (1 mg/mL). There is a large decrease in impedance between 1 layer (intersecting the x-axis at > 10^{11} Ω) and 4 layers (intersection the x-axis at <10^4), suggesting that percolation occurs at 4 layers. Additionally, the multiple cusps in the 4 layer data set indicate that there are multiple electronic processes occurring in the 4 layer film. The largest amount of information about conduction within MWNT thin films can be obtained from this graph and is discussed further in ref. 5. At 8 layers, the data simplifies to only a segment of the 4 layer data, and at 12 layers and beyond, the impedance data is represented by a straight line, suggesting bulk metallic CNT behavior in the films.

In order to determine the effect concentration has on the properties of films measured in-plane, 4 layer films were deposited using solution concentrations from 0.1 to 5 mg/mL. The impedance results for the varying concentration films measured in-plane are given in figure 1b. Analyzing these results in the same manner as before, a percolation event occurs between 0.5 and 1 mg/mL where there is the largest decrease in impedance. At 5 mg/mL, the impedance behavior again matches that for bulk metallic CNT films. The evolution of the in-plane impedance spectra with increasing solution concentration (figure 1b) seems to match the evolution of the in-plane impedance spectra with increasing number of layers (figure 1a), suggesting that, for surface properties, the same electronic behavior can be attained by varying either MWNT concentration or number of layers deposited.

Figure 2 shows SEM images corresponding to the in-plane for the films discussed both in the layer study (top row) and for the concentration study (bottom row), all taken at the same magnification. For the layer study, images for films containing 1 layer (2a), 4 layers (2b), 8 layers (2c), and 12 layers (2d) are presented. In figure 2a, the paper structure dominates the image with only a small area of the image showing interconnected MWNTs. However, in figure 2b, a fully connected MWNT film is shown, although the paper substrate can still be seen. In both cases, the entire area of the SEM images (figs 2c for 8 layers and 2d for 12 layers) shows a highly dense MWNT film structure. SEM images for the concentration study at 4 layers are shown for 0.1 mg/mL (2e), 0.5 mg/mL (2f), 1 mg/mL (2g), and 5 mg/mL (2h). The images for 4 layer films deposited from dispersions containing less than 1 mg/mL MWNT concentration (figs. 2e and 2f) show a similar structure to that seen in figure 2a, for the single layer film deposited using 1mg/ml. Figure 2g provides details for another region of the 4 layer film created from 1 mg/mL solution (figure 2b), showing an interconnected film and bridging of that film over the pore structure of the paper.

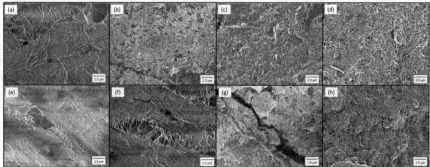

Figure 2: SEM micrographs for the in-plane layer study (top row) and concentration study (bottom row). Top: a) 1 layer; b) 4 layers; c) 8 layers; d) 12 layers. Bottom: e) 0.1 mg/mL; f) 0.5 mg/mL; g) 1 mg/mL; and h) 5 mg/mL. The scale bar reads 2.5 μm for all images.

Thru-Plane

Figure 1c gives the impedance data for the thru-plane layer study done at a constant concentration of MWNTs (1 mg/mL). There are two notable results from this data: 1) there is a drop in impedance between 8 and 12 layers (from 10^6 Ω to $\sim 10^4$ Ω); 2) the impedance at 1 layer in the thru plane is $\sim 2 \times 10^6$ Ω whereas in-plane it measures $\sim 10^{11}$ Ω (from figure 1a), suggesting that there is interconnection of MWNTs through the thickness of the paper at 1 layer deposition. This interconnection is controlled by the pore network of the paper, as current flows detected by C-AFM in regions similar in size to the pores (not shown) have been obtained in similarly made films[6].

Since deposition of less than one layer is not possible under constant concentration constraints, decreasing the solution concentration simulates a less than one layer deposition.

Figure 1d shows the impedance results for concentration variation of 1 layer films measured in the thru-plane. It is shown that between 0.5 and 1 mg/mL there is a 2 order of magnitude drop in the impedance, suggesting that percolation is happening between these two concentrations. Increasing the solution concentration from 1 mg/mL to 5 mg/mL leads to an order of magnitude increase in impedance, which is likely due to the large concentration of MWNTs present in the solution and the increased amount of surfactant needed to disperse them. Because film that is formed at such high concentrations is thick, the dense MWNT film likely resulted in increased surfactant retention in the film during filtration as well as less penetration of the MWNTs through the paper substrate due to bridging over pores rather than filling in the pores. Similarly, an increase in resistance is observed between 16 and 20 layers in the thru-plane, although the magnitude of the increase in this case is not as large. The percolation event which occurs between 8 and 12 layers is likely due to an increased number of interconnections between the MWNTs, increasing the number of connections from the surface to the back of the paper substrate.

Percolation Results

Percolation data for all the data sets discussed is summarized in figure 3. Figure 3(left) shows the results for the in-plane studies, plotting sheet resistance vs. number of layers or concentration. The square data points (layer study) correspond to the left and bottom axes, while the circle data points (concentration study) is plotted with respect to the the right and top axes. Likewise, figure 3(right) shows the percolation data for the thru-plane studies with a similar arrangement of data and axes as the in-plane studies, except that resistivity is being plotted rather than sheet resistance in this case. It is notable that the data for samples measured in-plane and thru-plane that correspond to the same conditions match up in both cases.

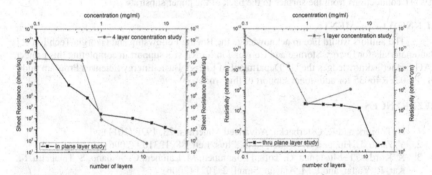

Figure 3: Percolation curves for all studies detailed in this paper. Left) In-plane study results: sheet resistance vs. number of layers (square data points) and sheet resistance vs. concentration (circle data points); Right) Thru-plane study results: resistivity vs. number of layers (square data points) and resistivity vs. concentration (circle data points).

CONCLUSIONS

It is possible to deposit highly conductive films of MWNTs on paper substrates from aqueous solution using a vacuum filtration-aided dropcasting technique. Impedance spectroscopy can be used to track how the electrical properties of the MWNT film change on the surface of the paper and within the thickness of the paper by varying the number of layers or the concentration of the MWNT dispersion. Additionally, the overall electrical response of the films can be extracted and used to determine percolation thresholds as a function of layer number or concentration. It is worth noting that these results were compared to 4-probe DC electrical measurements in order to confirm that the overall response extrapolated from the AC measurement data was accurate, as was done in Ref. 5. By understanding of the electrical properties which arise both in-plane and thru-plane in these composites, better control of the structure and electrical properties of the MWNT films can be obtained.

In the case of the in-plane measurements, varying either solution concentration or layer number results in the same shape of percolation curve as well as the same type of evolution of the impedance spectra determined. With respect to varying layer number (constant solution concentration = 1 mg/mL), the films percolate in-plane at 4 layers; with respect to concentration (constant layer number = 4), the films percolate at 1 mg/mL. Sheet resistances as low as 100 $\Omega/\square$ were obtained.

In the case of the thru-plane, at 1 mg/mL concentration, the MWNTs have already percolated through the paper structure at 1 layer with a resistivity of ~10^6 Ω-cm. The percolation threshold value in the thru-plane is high because the paper has a large contribution. By decreasing the concentration of the solution, sub-1-layer films can be simulated. When solution concentration is less than 1 mg/mL, the impedance response is very high (> 10^8 Ω), over 2 orders of magnitude higher than the resistivity at 1 mg/mL. As layer number is increased, another percolation event happens between 8 and 12 layers which is likely due to a larger number of MWNT connections from the surface to the back of the paper substrate.

ACKNOWLEDGMENTS

The authors would like to acknowledge the Boeing Fellowship and Georgia Tech IGERT: Nanomaterials for Energy Storage and Conversion for RLM's support in completing this project. RAG further acknowledges the US Department of Energy Basic Energy Sciences Program under DE-FG02ER46035 for additional support of this work.

REFERENCES

1. D. Tobjork and R. Osterbacka, Advanced Materials. 23, 1935 (2011).
2. Y. Zhou, L. Hu, and G. Gruner, Appl. Phys. Lett. 88, 123109 (2006).
3. K. Kordas, T. Mustonen, G. Toth, H. Jantunen, M. Lajunen, C. Soldano, S. Talapatra, S. Kar, R. Vajtai, and P.M. Ajayan, Small. 2, 1021 (2006).
4. R.H. Baughman, A.A. Zakhidov, and W.A. de Heer, Science. 297, 787 (2002).
5. R.L. Muhlbauer, S.M. Joshi, and R.A. Gerhardt, J. Mater. Res. 28, 1617 (2013).
6. R.L. Muhlbauer and R.A. Gerhardt, Appl.Phys. Lett., submitted January 2013.
7. M.P. Garrett, I.N. Ivanov, R.A. Gerhardt, A.A. Puretzky, and D.B. Geohegan. Appl. Phys. Lett. 97, 163105 (2010).

Mater. Res. Soc. Symp. Proc. Vol. 1549 © 2013 Materials Research Society
DOI: 10.1557/opl.2013.1033

The electric resistance and the transport properties of carbon nanotube with a Cu chain: A First-Principle study

Chengyu Yang and Quanfang Chen

Mechanical and Aerospace Engineering Department, University of Central Florida,
Orlando, Florida, 328162450, USA
Quanfang.Chen@ucf.edu

ABSTRACT The electric resistance and the transport properties of a carbon nanotube (5,5) adsorbed with a copper chain connected with two copper end electrodes have been calculated by employing the nonequilibrium Green's function and the Density Function Theory. The properties of the pure carbon nanotube (5,5) with the Cu electrodes have also been calculated as a reference. Both the equilibrium and the nonequilibrium conditions have been investigated. The results have shown that the electrical resistance of the metallic CNT (5,5) has been reduced by the adsorption of the Cu chain due to the interaction between the Cu and the CNT. The change of the I-V curve slope is also explained in terms of the transmission spectrum.

1. INTRODUCTION

Metallic carbon nanotube (CNT) has been proven to be ballistic in transport with a large electron mean free path and a large current carrying capability. As a result, it has long been regarded as a potential replacement for Cu interconnects in nanoelectronics . However, due to the low density of states of CNTs near the Fermi level, a single carbon nanotube has a intrinsic resistance of about 6.5 k [1], which is greater than copper and could cause excessive RC delays of signals. On the other hand, bundles of carbon nanotubes, since the parallel channels contribute to the conduction, have been proposed and experimentally proved as a better candidate than the individual CNT [24]. However, it is difficult to utilize carbon nanotubes alone as the interconnects in the actual device [5]. Metals on the other hand, could provide carbon nanotubes' both the support and the link with their environment. Therefore metal/carbon nanotube hybrid systems would be a most important system in nanotechnology [6].

Titanium chain has been proved to change Carbon nanotube's electronic structure and the conduction nature[7]. It has been reported that the incorporation of a titanium chain modifies the electronic structure of carbon nanotube via the charge transfer and the orbital hybridization[5, 8, 9]. In comparison to titanium, copper is also a transition metal with 3d electrons that could interact with carbon nanotubes. Our assumption is that Cu could generate delocalized states in utilizing carbon nanotube's long mean free path, thus produce a potential higher electric conductance in the form of hybrid material. Instead of using the parallel channels in CNT bundles, free electrons from Cu contribute to CNT's density of states (DOS) at the Fermi level. Therefore the Cu atoms may enhance the CNT's conductance and provide a mechanical support and the electronic link to its environment at the same time.

2. CALCULATION AND SIMULATION MODEL

In this article, we present the Cu chain's effect on the transport properties and the electric conductance of a carbon nanotube in forming the Cu/CNT/Cu junction. As shown in Figure 1a, the Cu/CNT/Cu junction has a CNT in the center, and two end contacting Cu electrodes on both sides. Similarly a copper chain is placed onto the CNT and linked to the two end electrodes to form the Cu/CNT+Cu/Cu system, as depicted in Figure 1b. CNT(5,5) is selected in all cases of this study. The length of the CNT(5,5) is more than 1 nm, or about six unit cells long. Each electrode consists of $5\times5\times4$ Cu atoms. The distance between the electrode and the central CNT has been optimized[10, 11] and the equilibrium relaxation distance is 1.71 Å.

Density function theory and the Nonequilibrium Green function are employed to calculate the transport properties of our systems. The general gradient approximation(GGA) and the PerdewBurkeErnzerhof (PBE) pseudo potentials with double zeta basis sets were used. A tolerance of 1×10^{5} has been used as the energetic convergence criterion. The k point samplings are set as $3\times3\times100$ for both systems (with/without the Cu chain).The commercial software QUANTUMWISE [13] has been used to perform all calculations.

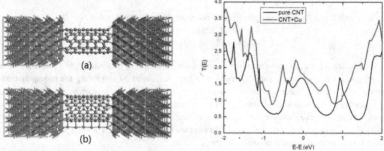

Figure 1 (a) The geometry sketch of the Cu/CNT(5,5)/Cu two probe system. (b) The geometry sketch of the Cu/CNT(5,5)+Cu/Cu two probe system.

Figure 2 Transmission spectrum of the Cu/CNT/Cu junction and the Cu/CNT + Cu/Cu junction

3. RESULTS AND DISCUSSION

3.1 Transport properties at equilibrium

3.1.1 Transmission spectrum at zero bias

The equilibrium conductance is related to the transmission coefficients T(E) under zero bias[12]. In order to compare the equilibrium conductance, the transmission coefficients at the zero bias for the two systems have been calculated and are presented in Figure 2. The black line in Figure 2 represents the transmission coefficient of the Cu/CNT/Cu system, while the red line shows the transmission coefficient of the Cu/CNT+Cu/Cu system. The transmission spectrum of the Cu/CNT(5,5)/Cu junction presented in Figure 2 agrees well with the reported results [10] by others, although there minor difference on the device setup.

Figure 2 shows that the addition of Cu chain has brought significant effect on the transmission coefficient of the Cu/CNT/Cu junction. In the vicinity of the Fermi level, the transmission coefficient of the Cu/CNT/Cu system has been increased, indicating that the conductance of the Cu/CNT/Cu system is also increased by incorporating the Cu chain.

The equilibrium conductance for the two systems could be calculated by the following equation[12, 14]:

$$G = \frac{2e^2}{h}T(E_f, V_b = 0V) \quad (1)$$

So the conductance of the two systems could be evaluated by using the transmission coefficients T at the Fermi level at the zero bias voltage. From the Figure 2 that the $T(E_f)$ of the Cu/CNT/Cu system is about 1.66304, while the $T(E_f)$ for the Cu/CNT+Cu/Cu system is about 1.779245. According to equation (1), we could get the conductance for each system, i.e., 128.9 µS for the Cu/CNT/Cu junction, and 137.908 µS for the Cu/CNT+Cu/Cu junction.

3.1.2 Current-voltage (I-V) curve of the two probe systems

The current that passes through the center region can be calculated by using the Landauer-Büttiker formula[12] :

$$I(V_b) = \frac{2e}{h} \int_{\mu_L}^{\mu_R} dE (f_L(E, V_b) - f_R(E, V_b)) T(E, V_b) \quad (2)$$

Where μ_L and μ_R are the chemical potential of the left and right electrodes, $f_L(E,V_b)$ and $f_R(E,V_b)$ are the Fermi-Dirac functions functions under bias V_b in the left and the right . The $T(E,V_b)$ is the transmission coefficient under bias V_b.

Figure 3 shows the current-voltage (I-V) curve as well as the slope of the I-V curve for the two probe systems. The Cu/CNT + Cu/Cu system shows a higher conductance than that of the Cu/CNT/Cu system. The I-V curves at low bias show a linear behavior (Figure 3a) from which we can use the Ohm law to calculate the conductance of the system. By averaging the data below 0.1V, we could calculate the conductance for the Cu/Cu+ CNT/Cu as 134.73 µS, while the conductance for the Cu/CNT/Cu is about 127.05 µS. The resistance calculated from the I-V curve didn't deviate much from the resistance we have obtained from the equilibrium condition.

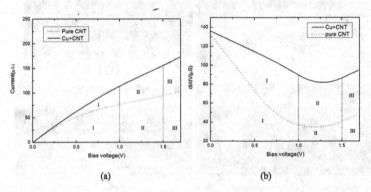

(a)   (b)

Figure 3 Current voltage (I-V) curve of the two probe systems. (a) I-V curve. (b) Slope of I-V curve as a function of bias voltage.

Table 1 shows the summary of the calculation results. Comparing the data we have achieved, one could conclude that the CNT+Cu hybrid nano wire has a greater conductance than that of the pure CNT at both equilibrium and low bias voltage cases, with the conductance enhanced by 7.1% and 5.7% respectively.

Table.1 The conductance of the two probe systems

		Cu/CNT(5,5)/Cu	Cu/CNT(5,5)+Cu/Cu	Increment
Equilibrium state	Conductance/µS	128.9	137.908	7.1%
Non-equilibrium state	Conductance/µS	127.05	134.73	5.7%

3.2 Transmission eigenstates at the Fermi level

Transmission eigenstates indicate the electronic states that contribute to the conductance [16]. By performing the analysis of eigenstates , we found that there're two main transmission eigenstates for the Cu/CNT/Cu system, and three transmission eigenstates for the Cu/CNT+Cu/Cu system. Since the whole central parts in both devices are metallic, we could just analysis the transmission eigenstates at the Fermi level. Figure 4a &Figure 4c present the primary transmission eigenstate for the Cu/CNT/Cu in the form of isosurface and contour, while Figure 4b

and Figure 4d present the primary transmission eigenstate for the Cu/CNT+Cu/Cu in the corresponding forms. From Figure 4, we found that the eigenstate for the pure CNT were distributed evenly over the whole nanotube, while the eigenstate for the CNT/Cu case were mainly concentrated in the region consists of Cu and the carbon nanotube. The interaction between the Cu and the CNT can be seen clearly from Figure 4b and Figure 4d. It is suggested that the electrons pass the system mainly via the interaction region between the Cu chain and the CNT. This interaction between Cu and CNT could be responsible for the increased conductance of the Cu/Cu+CNT/CNT system. Our analysis of all the other eigenstates denotes the similar pattern as the main eigenstate presented here.

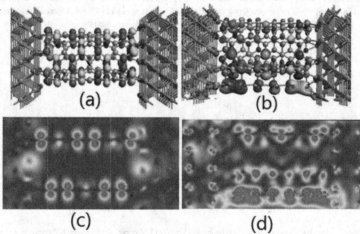

Figure 4 Transmission eigenstates at Fermi level: (a) isosurface for Cu/CNT/Cu junction. (b) isosurface for Cu/CNT+Cu/Cu junction. (c) contour for Cu/CNT/Cu junction. (d) contour for Cu/CNT+Cu/Cu junction.

3.3 Current density analysis

The current density gives us a direct picture of how the current flows under a small bias voltage in the conduction system. To intuitively observe how the current flows in the Cu/CNT+Cu/Cu system, we have plotted the current density in both isosurface and contour forms. In the isosurface image (Figure 5a), the shape of the isosurface shows the value of the current density in the corresponding area. We found that the current were mainly concentrated in the region consists of the Cu chain and the carbon nanotube, while in other regions the current density has a much less value. In the contour image for the Cu/CNT+Cu /Cu system, the cut plane was made by cutting through the Cu chain. The brighter the color of the cutplane, the stronger the current density flows in this area. Similarly, it is found that the current mainly flows through the region of Cu chain and the carbon nanotube at a small bias voltage.

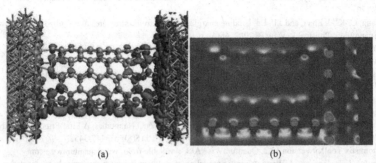

(a)   (b)

Figure 5 Current density at a small bias voltage: (a) isosurface for the Cu/CNT+Cu/Cu junction. (b) contour for the Cu/CNT+Cu/Cu junction.

4. CONCLUSIONS

Using the nonequilibrium Green's function in conjunction with DFT, we have obtained the currentvoltage characteristics for both the Cu/CNT/Cu junction and the Cu/CNT+Cu/CNT junction. Our calculations have shown that by incorporating Cu atoms onto Cu/CNT/Cu system, the system with an enhanced conductance under both equilibrium and nonequilibrium conditions has been achieved. The interaction between the Cu chain and the CNT could be recognized from the main transmission eigenstates and the current density plots. This interaction has increased the system's electric conductance by about 7%. The features explored here may benefit the future nano electronics and NEMS, since significantly increased conductance could be realized by incorperating metals like copper with carbon nanotubes.

ACKNOWLEDGEMENT

The author acknowledges the financial support from the National Science Foundation of the United State of America and the International Copper Association.

REFERENCES

1   Park, P.L.M.J.-Y., Electron Transport in Single-Walled Carbon Nanotubes. MRS BULLETIN, 2004.
2   Andreas Thess, R.L., Pavel Nikolaev,Hongjie Dai,Pierre Petit,Jerome Robert,Chunhui Xu, Young Hee Lee,Seong Gon Kim, Andrew G. Rinzler,Daniel T. Colbert,Gustavo E. Scuseria, David Tománek,John E. Fischer, Richard E. Smalley, Crystalline Ropes of Metallic Carbon Nanotubes. Science, 1996. 273(5274).
3   Jun Li, Q.Y., Alan Cassell, Hou Tee Ng, Ramsey Stevens et al., Bottom-up approach for carbon nanotube interconnects. Applied Physics Letters, 2003. 82(15): p. 2491.
4   Franz Kreupl, A.P.G., Maik Liebau, Georg S. Duesberg, Robert Seidel, Eugen Unger, Carbon Nanotubes for Interconnect Applications in Proc. IEEE Int. Electron Devices Meeting. 2004.
5   Youngmi Cho, C.K.H.M., Youngmin Choi, Sohee Park, Choong-Ki Lee, and Seungwu Han, Electronic Structure Tailoring and Selective Adsorption Mechanism of Metal-coated Nanotubes. Nano Letters 2007. 8(1): p.
6   Banhart, F., Interactions between metals and carbon nanotubes: at the interface between old and new

materials. Nanoscale, 2009. **1**(2): p. 201.

7   Durgun, E., et al., Systematic study of adsorption of single atoms on a carbon nanotube. Physical Review B, 2003. **67**(20).

8   Yang, C.-K., J. Zhao, and J.P. Lu, Binding energies and electronic structures of adsorbed titanium chains on carbon nanotubes. Physical Review B, 2002. **66**(4): p. 041403.

9   Yang, C.-K., Complete Spin Polarization for a Carbon Nanotube with an Adsorbed Atomic Transition-Metal Chain. Nano Letters, 2004. **4**(4).

10   Gao, F., J. Qu, and M. Yao, Effects of local structural defects on the electron transport in a carbon nanotube between Cu electrodes. Applied Physics Letters, 2010. **97**(24): p. 242112.

11   Gao, F., J. Qu, and M. Yao, Electronic structure and contact resistance at an open-end carbon nanotube and copper interface. Applied Physics Letters, 2010. **96**(10): p. 102108.

12   Han, Q., et al., Electrical Transport Study of Single-Walled ZnO Nanotubes: A First-Principles Study of the Length Dependence. The Journal of Physical Chemistry C, 2011. **115**(8): p. 3447-3452.

13   Atomistix ToolKit version 12.2. QuantumWise A/S; Available from: www.quantumwise.com.

14   Taylor, J., H. Guo, and J. Wang, Ab initio modeling of quantum transport properties of molecular electronic devices. Physical Review B, 2001. **63**(24): p. 245407.

15   Gao, F., J. Qu, and M. Yao, Electrical Contact Resistance at the Carbon Nanotube/Pd and Carbon Nanotube/Al Interfaces in End-Contact by First-Principles Calculations. Journal of Electronic Packaging, 2011. **133**(2): p. 020908-4.

16   Xu, B. and Y.P. Feng, Electronic structures and transport properties of sulfurized carbon nanotubes. Solid State Communications, 2010. **150**(41–42): p. 2015-2019.

Mater. Res. Soc. Symp. Proc. Vol. 1549 © 2013 Materials Research Society
DOI: 10.1557/opl.2013.1034

Field emission vacuum electronic devices utilizing ultrathin carbon nanotube sheet

Hai H. Van and Mei Zhang
Department of Industrial and Manufacturing Engineering, FAMU-FSU College of Engineering;
High-Performance Materials Institute, Florida State University, 2525 Pottsdamer Street,
Tallahassee, FL 32310, U.S.A.

ABSTRACT

Novel field emission (FE) devices are introduced employing lateral architecture. Ultrathin multiwalled carbon nanotube (MWCNT) sheet were utilized to fabricate the emitter. Effects of basic configuration of sheets, including the orientation of CNTs and sheet thickness were examined. The novel device achieved the threshold field (the electric field at which current density reach 1 mA/cm^2) of 0.67 V/µm and enhancement factor larger than 20,000.

INTRODUCTION

Electron FE is an emission of electrons induced by electrostatic field. The performance of an FE device depends on geometrical properties of its emitters, including: aspect ratio, uniformity, and alignment of emitters, distance between emitters, uniformity of the emitter array, and interelectrode distance. In addition, the material that has lower work function and sharper tip is suitable as an electron emitter. CNTs are considered as one of such materials because of their nanometer size tips, high aspect ratios, high electrical and thermal conductivity, high mechanical strength, and high chemical stability.

The common architecture of field emission device is vertical architecture [1], in which 2 electrodes are set parallel to each other. The emitter arrays are distributed on the surface of one electrode (i.e. cathode), facing the other electrode (anode) as illustrated in Figure 1a. The issue with this architecture is that because the emitters are arranged closely to each other, the screening effect imposes heavily on the emitters. According to the simulation model Figure 1b, the field energy is not uniformly distributed. It is much stronger at the edge (namely the edge effect), and weaker in the central area [2]. The non-uniform distribution of field energy will reduce the total performance of the device. Thus, if the size of emitter array is reduced into one dimension, the screening effect is limited. In addition, the edge effect will be taken advantage.

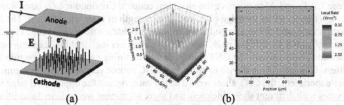

Figure 1: (a) FE device with vertical architecture, (b) simulation model of field energy distribution of vertical FE device [2].

Based on this analysis, FE lateral structure is utilized. In this structure, both electrodes are set in-plane together with the emitter array, as described in

Figure 2. This architecture has the benefits of relieving the screening effect in one dimension, together with the small size, flexible and readily to be integrated into electronic system. To achieve those benefits, the ultrathin freestanding MWCNT sheets are used as the base for emitter fabrication. Details relating to the sheet structure will be described in next section.

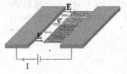

Figure 2: FE device with lateral architecture.

EXPERIMENT

The CNT sheet was produced by laterally drawing CNTs from a CNT forest synthesized by chemical vapor deposition (CVD). The CNT sheet was freestanding with the thickness less than 100 nm, which is suitable for lateral architecture. On the other hand, the CNTs were uniformly aligned and distributed in the sheet, which was perfect to function as FE emitters.

Figure 3: SEM image of CNT sheet.

A pair of aluminum (Al) rods was used as the holder for the sheets. Sheets continuously drawn from CNT forest were stacked on each other on the holder with certain angle relative to CNTs' alignment. The stacked structure was then densified in the Isopropyl Alcohol (IPA) and dried in vacuum at 150°C in 2 hours. This densification process improved the mechanical robustness, reduced the porosity, and enhanced the electrical conductivity. The thickness of 4-layer densified sheet was about 210 nm, measured by AFM. Silver paste was applied at the contact between CNTs and Al to improve the electrical contact. Continuous CO_2 laser was utilized to cut CNT sheet along the edge of Al rod. Wavelength of laser was 10.6μm. The laser power, irradiation time, and the scan speed were adjusted to obtain effective cutting. Typically, the power was kept at 0.5 W with the pulse rate of 500 pulses per inch and the scanning rate of 2.54 cm per second. As the result, the CNTs cut by laser remaining on the sheets functioned as the field emitters. Connecting the device to appropriate DC source generated the field emission current. In this experiment, the CNT sheet workd as the cathode and the opposite Al edge works as anode. Devices with different sheet thickness and sheet alignment were made. In addition, MWCNT buckypaper was made to test the effect of alignment and thickness of the sheet.

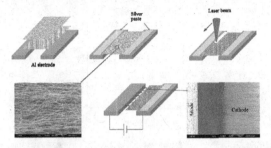

Figure 4: Process of CNT FE lateral device fabrication.

DISCUSSION

SEM and TEM images provide more details about the CNT emitters produced by this method. Firstly, the fabricated emitters were uniform in term of the alignment and the length of emitters. Secondly, the CNTs synthesized by CVD were MWCNTs with 4-5 walls (Figure 5b). Under the irradiation of laser, emitter structure was modified into double-walled CNTs (Figure 5c). In addition, thermal effect bundled CNTs together, making it well separate, which further reduced the screening effect.

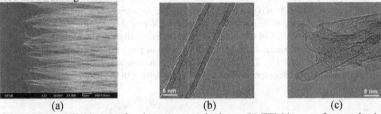

(a)   (b)   (c)

Figure 5: (a) SEM image of emitter array cut by laser, (b) TEM image of as-synthesized MWCNTs with 5 walls, (c) TEM image of CNT tip with 2 graphitic walls.

To verify the effect of the thickness on field emission performance, the devices were made with sheets stacked by 4 and 12 layers CNT sheets, and with buckypaper. The buckypaper was made by vacuum filtration of MWCNT suspension. The CNTs in the buckypaper were randomly distributed. The buckypaper was about 9 μm thick. The 4-layer and 12 layer CNT emitters did not show much difference although they were with different thicknesses (Figure 6a). The threshold fields corresponding to the device with the sheet thickness of 4 layers (210 nm), 12 layers, and with buckypaper (9 μm) are 0.67 V/μm, 0.68 V/μm, and 1.22 V/μm, respectively. The 4-layer and 12-layer sheet devices performed the field enhancement factor of 21687 and 20945, respectively. The reason was because after the laser cutting process, nanotubes along the edge of the sheet were bundled together (Figure 6b). Thus, the difference of number of layers did not affect the thickness of emitter array. On the other hand, buckypaper showed less effective emission behavior. Observing the SEM images shows that the edge of buckypaper shows layers of emitters, which are randomly aligned. Thus, the screening effect of buckypaper was worse than that of stacked CNT sheets.

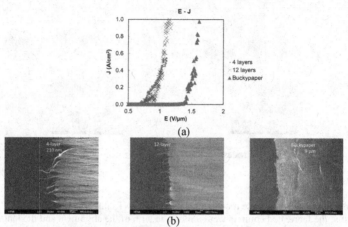

(a)

(b)

Figure 6: FE devices utilizing CNT sheets with different thicknesses: (a) current density performance, (b) SEM images of corresponding sheets.

CNT sheets with different orientations were fabricated. Figure 7a shows that the more aligned CNTs to emission direction result in the better FE performance. The threshold fields corresponding to the orientation of 5°, 40°, and 60° are 0.67 V/µm, 0.86 V/µm, and 0.98 V/µm, respectively. Different alignments of the sheets created different alignment and distribution of emitter array. Under the electrostatic field, the emitters were aligned towards anode, which was parallel to the electrostatic force [3]. The smaller angles that nanotubes align relatively to the emission direction make that phenomenon less effective, resulting in the better FE performance.

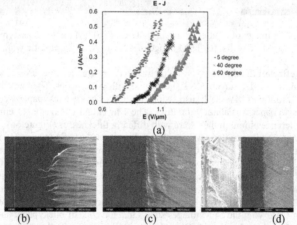

(a)

(b)   (c)   (d)

Figure 7: FE devices utilizing CNT sheets with different orientations: (a) current density performance, (b) SEM images of corresponding sheets.

The change of sheet structures over the time was tested with different levels of current density (Figure 8a). Figure 8b shows SEM image of the sheet before FE test. The result of initial FE measurement is shown as 1st plot in Figure 8a. The threshold field was 0.8 V/μm and current density was 0.8 A/cm^2 at 1.38 V/μm. The 2nd plot showed FE result after 20 times of FE tests. At this point, the threshold field increased to 0.93 V/μm and current density was decreased to 0.22 V/μm at 1.38 V/μm. The SEM image described the change of the CNT arrangement at the edge of the sheet (Figure 8c). The CNTs started to be extracted from the structure and damaged. The emission current was increased to higher value 1.08 A/cm^2 at 1.7 V/μm (3rd plot). The SEM image after this measurement did not clearly show the change of the structure (Figure 8d). However, the next measurement (4th plot) showed that the threshold field was increased to 1.38 V/μm. At this time, the emission current was increased to 3.84 A/cm^2 at 2.41 V/μm. The structure was damaged by the high current density, which can be apparently observed in Figure 8e. These results prove that the structural change of emitter arrays starts from the tips of emitters and uniformly along the edge of the sheet which means the uniform emission of the CNT array.

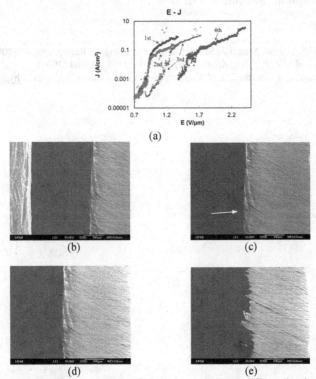

(a)

(b)

(c)

(d)

(e)

Figure 8: Performance change of FE device over time at different levels of current density: (a) FE performance, (b) (c) (d) (e) SEM images of emitter array structure corresponding to 1st, 2nd, 3rd, and 4th, respectively.

CONCLUSIONS

Lateral architecture and free standing CNT sheet are employed to make FE devices. The sheets present the ultrathin structure and uniformly aligned and distributed CNTs. The novel geometrical characteristics are beneficial to relieve screening effect in one dimension. Additionally, the sheets show the uniform electron emission along the emitter array. Device is fabricated by simple process. Several device configurations have been tested. Laser method helps to further reduce the screening effect. The threshold field was achieved at 0.67 V/μm with enhancement factor larger than 20,000. In addition, this architecture has the benefit of small size, flexibility, and possibility to be integrated into electronic system.

ACKNOWLEDGEMENTS

The authors would like to acknowledge the supports of High-Performance Materials Institute of Florida State University.

REFERENCES

1. N. S. Xu and S. E. Huq, Materials Science and Engineering: R: Reports, 48, 47 (2005).
2. R. C. Smith and S. R. P. Silva, Applied Physics Letters, 94, 133104 (2009).
3. G. S. Bocharov, a. a. Knizhnik, a. V. Eletskii, and T. J. Sommerer, Technical Physics, 57, 270 (2012).

Mater. Res. Soc. Symp. Proc. Vol. 1549 © 2013 Materials Research Society
DOI: 10.1557/opl.2013.685

Functionalized Carbon Nanotube Matrix for Inducing Noncovalent Interactions Toward Enhanced Catalytic Performance of Metallic Electrode

Le Q. Hoa,[1,2] Hiroyuki Yoshikawa,[1,2] Masato Saito[1] and Eiichi Tamiya[1]
[1] JST-CREST, 2-1 Yamadaoka, Suita, Osaka 565-0871, Japan
[2] Department of Applied Physics, Graduate School of Engineering, Osaka University,
2-1 Yamadaoka, Suita, Osaka 565-0871, Japan

ABSTRACT

Herein, we investigated the noncovalent interactions derived from functionalized carbon nanotube matrices grafting metallic alloy PtRu nanoparticle-decorated Vulcan carbon (fMWCNTs-g-PtRu/C) toward the enhancement of alcohol oxidation reactions. The fMWCNTs noncovalently grafted PtRu/C was successfully synthesized and demonstrated significant enhancement of the electro-catalytic activity and stability toward alcohol oxidation reactions as revealed by electrochemical characterizations. The presented results indicate that the grafting matrix specifically enhances ethanol oxidation reaction kinetics much more than methanol and propanol oxidation reactions. Since the same loading of PtRu/C was used for all tests, the differentiation between these reactants revealed the different strength of noncovalent interactions between the functional matrix and corresponding reactants. This result reveals a new strategy for using fMWCNTs matrix as potential catalyst supports due to its facile fabrication and functionalization, cost effectiveness and environmental friendliness, factors in which all of them are necessary for the practical application of direct alcohol fuel cells in near future.

INTRODUCTION

The utilization of direct alcohol fuel cells (DAFCs) technology as a viable green energy has been hindered because of the sluggish kinetics of the inefficient alcohol oxidation reactions, low durability of catalysts, difficulties in fuel cell fabrication procedures and cost effectiveness for large-scale implementation [1, 2].

It is well-established that noble metals (e.g. Pt and Au) induce efficient adsorption and dis-sociation of small organic molecules, rendering them excellent catalysts in both anode and cathode fuel cell reactions. However, such noble metals are very expensive. To reduce their cost, the conventional approach is to combine noble metals with other transition metals/metal oxides to produce bimetallic, and ternary metallic alloys. Unfortunately, the cost of transition metals for successful application, in many cases, is actually higher than that of noble metals (e.g. Rh in $PtRhSnO_2$) [3, 4]. Despite intensive theoretical and experimental research efforts in searching for high performance low-cost materials, research into the role of matrices and solvents in oxidation reaction pathways is relatively new. Thus, further application steps in tuning electrocatalytic activities by functional matrices are under exploration.

Differing from the conventional inorganic-based approach, we are utilizing carbon-based materials as functional supporting matrices to enhance the performance of low-loaded noble metal-based catalysts [5-9]. In our previous work, we explored effects of functionalized multiwalled carbon nanotubes (fMWCNTs) toward ethanol oxidation reaction on Pt nano particles by two distinct assembly structure: multilayered and grafted [8, 9]. We discovered that the fMWCNTs "layer" or "shell" can induce non-covalent interactions between functional groups and the Pt active sites with the reactant and the intermediates that promote reaction kinetics via stabilizing effects. For the different functional groups, the various noncovalent interactions [10, 11], which include hydrogen bonding of functional group-water, functional group-ethanol, and functional group-OH_{ads}-Pt bonding, are presented between the Pt surface and the electrolytic species. These interactions vary depending on nature of functional groups, leading to significant variation in the electrocatalytic reaction rates. The ethanol oxidation catalytic activity and stability were enhanced as much as 100% by the grafted structure compared to the multilayered assembly. In this study, as a next step of these works, we further investigated the catalytic activities and stability of fMWCNTs grafted commercialized bimetallic PtRu on Vulcan carbon (PtRu/C) toward various alcohol oxidation reactions (methanol, ethanol and 1-propanol) to reveal the impact of non-covalent interactions between functional matrices on PtRu active surface on different targets.

EXPERIMENT DETAILS

Reagent grade concentrated nitric acid, sulphuric acid, multi-walled carbon nanotubes (MWCNTs, purity ≥ 95%, 40-70 nm in diameter) were obtained from Wako Co., Japan. The electrocatalyst PtRu/C were purchased from FuelCellStore (HP 20% Pt:Ru on Vulcan XC-72). MWCNTs were surface oxidized by refluxing in conc. HNO_3 solution at 140 °C – 160 °C for 24 h, then filtered and dried to obtain MWCNTs-COOH. The MWCNTs-COOH-grafted-PtRu/C catalyst was synthesized by mixing and sonicating MWCNTs-COOH (5 mg) with PtRu/C (2.5 mg) in ethanol (99.5 %, 2 mL) and nafion (50 μL of 5%, Wako Co., Japan) followed by casting on a glassy carbon electrode (surface area 0.0706 cm^2) or carbon paper (geometric surface area 1cm^2, TMIL. Co. Japan) to obtain a PtRu loading of 0.025 mg cm^{-2}. The modified carbon paper electrode was then subjected to thermal treatment at 80 °C to remove the solvent and to increase the stability of the catalysts on carbon paper substrates. The surface morphologies of catalytic materials were acquired by scanning electron microscopy (SEM) (DB 235 microscope, FEI). Electrochemical measurements, including cyclic voltammetry (CV), chronoamperometry (CA), and electrochemical impedance spectroscopy (EIS) were performed with an Autolab potentiostat/galvanostat PGSTAT12 (EcoChemie, Netherlands). A three-electrode cell was used for the electrochemical measurement. A platinum rod served as the counter electrode and all potentials were measured with respect to an Ag/AgCl reference electrode. Prior to each measurement, the electrolyte was purged with nitrogen for 30 min, and the catalytic materials, either on glassy carbon or on carbon paper, were subjected to pre-treatment in the electrolyte (without reactant) by CV scanning 20 times in the potential window -0.25 V to 1.2 V (vs Ag/AgCl).

DISCUSSION

Electrochemical surface area characterization of PtRu with and without functionalized catalytic noncovalent grafting matrix

In brief, a grafted-catalytic structure was formed by thoroughly mixing and sonicating a solution containing both PtRu/C and MWCNTs-COOH, and then depositing on carbon substrate. The assembly process is thought to be through self-assembly via electrostatic (from charged functional groups -COO⁻) and hydrophobic interactions (from CNT backbone) toward PtRu surface. The resulting structure therefore features the powdery PtRu/C surrounded by interpenetrating MWCNTs-COOH network (figure 1A and 1B).

To reveal the difference in catalytic morphological structure, we carried out the electrochemical active surface area measurement that conventionally used to quantify the Pt active surface.

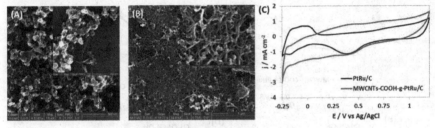

Figure 1. SEM images of PtRu/C (A) and MWCNTs-COOH grafted PtRu/C (B). (C) Electrochemical surface area characterization of PtRu in each catalytic structure using cyclic voltammetry. Scan rate is 50 mV s⁻¹, the electrolyte is H_2SO_4 0.5 M. PtRu loading in all cases is 0.025 mg cm⁻². All the catalysts were cast on pre-treated glassy carbon electrodes.

As shown in figure 1C, the hydrogen adsorption/desorption regions (from -0.25 V to 0.1 V) in the PtRu/C voltammogram could be clearly distinguished while it was diminished in the case of MWCNTs-COOH-g-PtRu/C. The reason for the larger hydrogen adsorption/desorption area of PtRu/C might be direct exposure of PtRu NPs to the electrolyte, in contrast to the grafted structures in which the PtRu surface was surrounded by MWCNTs-COOH. Since the same loading of PtRu/C was used, this result indicates that these electrochemical changes stem from different structural conformations adopted by the catalysts as influenced by the functional groups on the matrix. In the case of COOH-MWCNTs-g-PtRu/C, the out-of-pattern CV suggests that the COOH-MWCNTs may have totally buried the PtRu NPs, which is in line with aforementioned suggestion. Since this process is self-assembly, we further suggest that the MWCNTs-COOH can induce certain non-covalent interactions that maintain a specific active cavity that may suitable for a certain molecular size of reactant and intermediate. To investigate this possibility, we increased the size of reactant from 1 carbon (methanol) to 3 carbons (1-propanol) and studied the oxidation reactions happened within this grafted structure in the following sections.

Catalytic performance

In order to assess the enhanced performance of PtRu/C by grafting with functionalized MWCNTs toward ethanol oxidation reaction, we first carried out the electrochemical measurements to compare the catalytic activity and stability of MWCNTs-COOH-g-PtRu/C with that of PtRu/C on either glassy carbon (Fig. 2A) as ideal substrate or carbon paper (Fig. 2B) as realistic substrate. On glassy carbon electrode, although PtRu/C demonstrated earlier onset potential with higher current density, the oxidation peak actually ~100 mV positively shifted compared to that of the MWCNTs-COOH-g-PtRu/C clearly proving the important role of functionalized carbon nanotube matrix in improving PtRu catalytic activities. Especially, the sharper oxidation peak which attained higher current density in the case of COOH-MWCNTs grafted one, demonstrates the faster oxidation kinetics upon the presence of –COOH functional matrix.

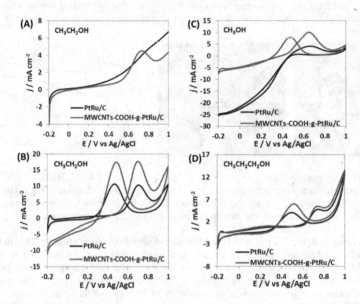

Figure 2. The electro-catalytic activities of MWCNTs-COOH grafted PtRu/C in comparison with native PtRu/C toward alcohol oxidation reaction. (A) The first anodic sweeps of PtRu/C and MWCNTs-COOH-g-PtRu/C on glassy carbon substrate in 1 M C_2H_5OH and 0.5 M H_2SO_4 at scan rate of 20 mV s^{-1}. (B) (C) and (D) are the 5th CVs of PtRu/C and MWCNTs-COOH-g-PtRu/C on carbon paper substrate in (B) 1 M C_2H_5OH, (C) 1 M CH_3OH, (D) 1 M $CH_3CH_2CH_2OH$ and 0.5 M H_2SO_4 at scan rate of 20 mV s^{-1}.

On carbon paper, however, the difference in oxidation peak position decreased ~10 mV but the current density of the grafted structure was 1.8 times higher than that of PtRu/C without functional matrix (figure 2B). With smaller reactant like methanol, the grafted structure exhibited 2 times higher current density than that of PtRu/C with earlier onset potential (figure 2C).

However, when the reactant contains up to 3 carbon, the differences in onset potential, oxidation peak position and current density were reduced significantly. It indicates that the active cavity induced by MWCNTs-COOH toward PtRu/C surface only be able to facilitate and stabilize small molecules, in particularly less-than-three-carbon reactant (figure 2D).

Stability performance

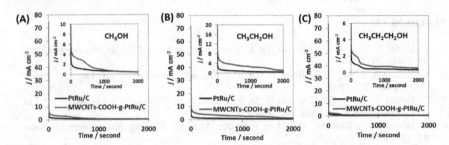

Figure 3. Effect of different reactant size on the stability performance of MWCNTs-COOH-g-PtRu/C in comparison with PtRu/C without functional matrix. Chronoamperometry measurements were carried out at 0.65 V (A), 0.7 V (B) and 0.75 (C) applied potentials in 1 M methanol, 1 M ethanol and 1 M 1-propanol in 0.5 M H_2SO_4, respectively. The insets show the magnification of the same data.

It has been clearly demonstrated by the CV data that (i) the catalytic activities of MWCNTs-COOH grafted PtRu/C is superior to that of PtRu/C without functional matrix and (ii) the degree of enhancement decreases when the size of reactant molecule increases. To further investigate how the grafted functional matrix stabilizes the size-varying intermediates during long time of oxidation reactions, we employed chronoamperometry at the oxidation potentials of each reactant. As shown in figure 3, for the first 500 seconds, in terms of oxidation current density enhancement, the following order was observed: methanol > ethanol > 1-propanol, which is in line with the catalytic activity trend analyzed before. However, from 500 seconds, the oxidation current density quickly decreased in the case of methanol and diminished in the case of 1-propanol, while better maintained in the case of ethanol. These differences suggest that the MWCNTs-COOH significantly stabilized the intermediates in the case of ethanol oxidation reaction but insufficiently stabilized intermediates in methanol and 1-propanol oxidation reactions.

CONCLUSIONS

In summary, the MWCNTs-COOH grafted PtRu/C was successfully synthesized and demonstrated significant enhancement of the electro-catalytic activities and stability toward the alcohol oxidation reactions. The enhancement was attributed to the noncovalent interactions between functional groups on the matrix and PtRu active site that help to facilitate the reactant and stabilize the intermediate species. The presented results further indicate that the as-prepared MWCNTs-COOH-g-Pt/Ru/C induced much more significant interactions toward ethanol and its intermediates than in the case of methanol and 1-propanol oxidation reactions. Detailed studies on the mechanistic aspects and further structural and functional optimizations toward specific targets are now in progress.

ACKNOWLEDGMENTS

This work was supported by JST-CREST.

REFERENCES

1. C. Lamy, S. Rousseau, E. M. Belgsir, C. Coutanceau and J. -M. Léger, *Electrochim. Acta* **49**, 3906 (2004).
2. C. Lamy, A. Lima, V. L. Rhun, F. Delime, C. Coutanceau and J. -M. Léger, *J. Power Sources* **105**, 285 (2002).
3. A. Kowal, M. Li, M. Shao, K. Sasaki, M. B. Vukmirovic, J. Zhang, N.S. Marinkovic, P. Liu, A. I. Frenkel and R. R. Adzic, *Nat. Mater.* **8**, 328 (2009).
4. A. Kowal, S. Lj. Gojkovic, K. S. Lee, P. Olszewski and Y. E. Sung, *Electrochem. Commun.* **11**, 724 (2009).
5. L. Q. Hoa, Y. Sugano, H. Yoshikawa, M. Saito, and E. Tamiya, *Biosens. Bioelectron.* **25**, 2509 (2010).
6. L. Q. Hoa, Y. Sugano, H. Yoshikawa, M. Saito, and E. Tamiya, *Electrochim. Acta* **56**, 9875 (2011).
7. L. Q. Hoa, H. Yoshikawa, M. Saito, and E. Tamiya, *J. Mater. Chem.* **21**, 4068 (2011).
8. L. Q. Hoa, M. C. Vestergaard, H. Yoshikawa, M. Saito, and E. Tamiya, *Electrochem. Commun.* **13**, 746 (2011).
9. L. Q. Hoa, M. C. Vestergaard, H. Yoshikawa, M. Saito, and E. Tamiya, *J. Mater. Chem.* **22**, 14705 (2012).
10. D. Strmcnik, K. Kodama, D. van der Vliet, J. Greeley, V. R. Stamenkovic and N. M. Marković, *Nat. Chem.* **1**, 466 (2009).
11. K. Müller-Dethlefs and P. Hobza, *Chem. Rev.* **100**, 143 (1999).

Carbon Nanomaterials

Mater. Res. Soc. Symp. Proc. Vol. 1549 © 2013 Materials Research Society
DOI: 10.1557/opl.2013.958

Reactive Template-Induced Self-Assembly to Ordered Mesoporous Polymer and Carbon

Yeru Liang, Ruowen Fu and Dingcai Wu[*]
Materials Science Institute, PCFM Lab and DSAPM Lab, School of Chemistry and Chemical Engineering, Sun Yat-sen University, Guangzhou 510275, P. R. China

ABSTRACT

As an important method for preparing ordered mesoporous polymer and carbon, organic template directed self-assembly is facing challenges because of the weak non-covalent interactions between the organic templates and the building blocks. Herein we developed a novel synthetic procedure based on a reactive template-induced self-assembly to construct ordered mesoporous framework. The aldehyde end-group of reactive template can react with the building blocks (i.e., resol) to form a stable covalent bond during the self-assembly process. This leads to an enhanced interaction between resol and template and thus achieves the formation of ordered mesostructure.

INTRODUCTION

Construction of well-defined porous materials has always been one of the most active areas in materials science, not only for its fundamental scientific interest, but also for many modern-day technological applications.[1] Significant progress has been attained in structural, compositional, and topological control in both inorganic and organic nanoporous structured materials. In recent years, there has been a growing focus on the synthesis and examination of porous materials with ordered mesopore array due to their unique characteristics including highly periodic pore arrangement, uniform mesopore size, and high surface area.[2, 3] These features are of great interest for a broad spectrum of applications, including energy storage, absorption, separation, drug delivery and catalysis.[4-7]

In general, organic template directed self-assembly is one of the most promising approaches toward synthesis of ordered mesoporous polymeric and carbonaceous materials.[8-10] Those organic compounds like amphiphilic surfactants and block copolymers, which can self-organize into a diversity of supermolecular structures, are generally used as pore templates. Cooperative self-assembly can drive building blocks around the supermolecular structures to form well-defined organic-organic mesophase. Direct removal of the organic templates by calcination or extraction with solvents leads to mesoporous framework. This organic template pathway has offered a powerful route to fabrication of ordered mesoporous polymer (OMP) and carbon materials with various mesostructures, suitable shapes, and designed functionalities. Despite these developments, however, challenges still remain, considering that in most cases, these ordered mesoporous products are not easily obtained by using the common organic template synthetic techniques. The main reason could be ascribed to the fact that organic templates usually interact with the building blocks through non-covalent interactions such as hydrogen bonding, van der Waals forces, and electrostatic interaction, which could be too weak to act as a driving force for formation of ordered mesostructures.

In this report, we describe a new strategy to address this shortcoming, which is based on the in-situ formation of strong covalent interactions during self-assembly. A reactive template is designed and introduced to induce a self-assembly with phenol/formaldehyde (PF) resol for construction of ordered mesoporous framework. This new proof-of-concept synthetic methodology could greatly develop the self-assembly theories for the fabrication of well-defined porous polymer and carbon materials.

The reactive template is synthesized by transformation of the chain ends of the commercial triblock copolymer EO_{106}-PO_{70}-EO_{106} (F127) from hydroxymethyl group to aldehyde group. The aldehyde end-group of this reactive F127 (R-F127) can react with PF resol to form a stable covalent bond during the self-assembly process. In this way, the greatly enhanced interaction resulting from a magnificent combination of the covalent bond and the hydrogen bond between

Figure 1. Schematic diagram for the route of reactive template-induced self-assembly to prepare the ordered mesoporous material.

the PF resol and the R-F127 leads to a successful self-assembly for formation of an ordered body-centered cubic mesostructure, thus obtaining OMP after template removal and OMC after carbonization (Figure 1).

EXPERIMENT

Preparation of the reactive template R-F127: The reactive template R-F127 containing the aldehyde end-group was synthesized by oxidization of the triblock copolymer F127 with the mixture of acetic anhydride (Ac_2O) and dimethyl sulfoxide (DMSO). 2.5 g of F127 was immersed in 10 ml of DMSO in a conical flask. Subsequently, 0.5 ml of Ac_2O was added to the mixture to achieve a final Ac_2O:-OH molar ratio of about 20:1. The reaction was allowed to proceed for about 8 h at room temperature under stirring. The resulting polymer sample was precipitated in 8 volumes of diethyl ether. After that, the polymer was redissolved in minimum methylene chloride and precipitated again in 8 volumes of diethyl ether. The above precipitation-dissolution procedure was repeated four times. The polymer was finally dried overnight under vacuum, leading to the R-F127.

Preparation of ordered mesoporous materials: 1.0 g of phenol and 2.8 ml of 40 wt% formaldehyde were dissolved in 25 ml of 0.1 M NaOH solution and then a clear phenol/formaldehyde resol was obtained after stirring at 68 °C for 30 minutes. Subsequently, 1 g of R-F127 was completely dissolved in 25 ml of water, followed by addition of the as-made resol solution. The mixture was continuously stirred at 65 °C for 96 h and then stirred at 70 °C for another 24 h. The product was collected, washed with water, and dried. After that, the as-made sample was calcined under N_2 flow at 350 °C for 3 h to decompose the triblock copolymer template and obtain the OMP. Further increasing the carbonization temperature to 800 °C led to formation of the OMC. For comparison, a control sample was synthesized. Its preparation procedure is exactly the same as that of the OMP except that the unreactive F127 was employed as the organic template.

RESULTS AND DISCUSSION

The as-prepared R-F127 is employed as the reactive template for cooperative self-assembly with PF resol in an aqueous solution. The resulting OMP sample shows three well-resolved diffraction peaks with a d spacing ratio of $1/(1/\sqrt{2})/(1/\sqrt{3})$ in its low-angle X-ray diffraction (XRD) patterns, which can be indexed as 110, 200, and 211 diffractions of a highly ordered body-centered cubic mesostructure (Figure 2). After carbonization at 800 °C in nitrogen flow, these three peaks still exist in the resulting XRD pattern, indicating that the ordered mesostructure is thermally stable (Figure 2). For comparison, control samples are prepared using the unreactive F127 as the template. As shown in Figure 2, the control samples do not display any diffraction peaks in the XRD pattern, indicative of the absence of mesostructure regularity. This is probably due to the weak interaction between F127 and PF resol. Such a comparison confirms that the reactive nature of R-F127 plays a crucial role in the cooperative assembly of ordered mesostructures.

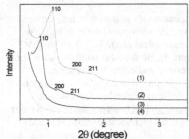

Figure 2. XRD patterns of typical samples.
(1). OMC; (2). OMP; (3). Control polymer sample;
(4). Control carbon sample

Transmission electron microscopy (TEM) images of the OMP and its corresponding control polymer sample are shown in Figure 3. The OMP reveals typical patterns of a body-centered cubic structure that contains well-distributed mesopores of about 5 nm, further confirming the successful formation of a cubic $Im\bar{3}m$ mesostructure . In contrast, the TEM characterization of the control sample shows a very poor and disordered pore framework. These findings are in good agreement with the XRD data shown in Figure 2. Taking the above results together, it is clearly demonstrated that the employment of the reactive template R-F127 is the key to formation of an ordered mesostructure. Generally, a periodic polymeric mesostructure only can be formed when there exists a strong enough interaction between building blocks (*e.g.*, polymer precursor) and organic template (*e.g.*, triblock copolymer). However, at present, the organic templates generally interact with the building blocks just through non-covalent interactions such as hydrogen bonding, van der Waals forces, and electrostatic interaction. Such kinds of interaction are too weak to act as a driving force for the formation of ordered polymeric mesostructures under many cooperative self-assembly conditions. For example, hydrogen-bonding interaction between the hydrophilic PEO segments of the unreactive F127 and hydroxyl group of PF resol is not large enough for the ordered assembly, resulting in a disordered mesostructure. When using the reactive template R-F127, the aldehyde groups of R-F127 can react with PF resol to form a stable covalent bond during the aqueous self-assemble process. Such an additional but necessary covalent bond greatly enhances the interaction between PF resol and template and thus allows for a successful self-assembly for an ordered mesostructure.

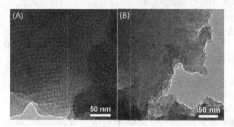

Figure 3. Typical TEM images of (A) OMP
and (B) the control polymer sample.

The pore structure of the as-prepared samples is quantitatively analyzed by measurement of N_2 adsorption at 77 K. Both OMP and OMC show a type IV adsorption isotherm, indicative of a uniform mesopore structure. The diameters of mesopore are calculated to be 5.0 nm for OMP and 3.2 nm for OMC with density functional theory (Figure 4). On the other hand, the adsorption amount of OMP increases very sharply at low relative pressure, indicating the existence of numerous micropores. These micropores mainly center at 1.3 nm. After carbonization, numerous new micropores of 0.5 nm diameter are generated within the resulting carbon framework of OMC probably due to burn-off of many non-carbon elements and carbon-containing compounds during pyrolysis (Figure 4). As a result, OMP and OMC exhibit high surface areas (311~637 m^2/g) and large pore volumes (0.25~0.32 cm^3/g).

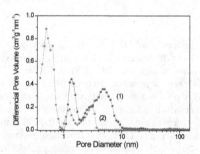

Figure 4. Pore size distributions of OMP and OMC.
(1). OMP; (2). OMC

CONCLUSIONS

In conclusion, a novel synthetic methodology based on the reactive template-induced self-assembly has been successfully developed to prepare ordered mesoporous polymeric and carbonaceous materials. The aldehyde group at the chain end of reactive template R-F127 can react with PF resol to *in-situ* form a stable covalent bond during the self-assembly process. With the employment of this reactive template for enhancing the interaction between PF resol and triblock copolymer, highly ordered body-centered cubic mesoporous polymer can be prepared. Such an ordered polymeric mesostructure is thermally stable and can be directly transformed into ordered mesoporous carbon after a carbonization process. We hope that this new concept of reactive template-induced self-assembly would promote the progress of fabrication science for mesoporous polymer and carbon materials

ACKNOWLEDGMENTS

We acknowledge financial support from the project of NSFC (51173213, 51172290, 50802116, 51232005), the Fundamental Research Funds for the Central Universities (09lgpy18), and the Project of Demonstration Base of Department of Education of Guangdong Province (cgzhzd0901).

REFERENCES

1. Wu, D.; Xu, F.; Sun, B.; Fu, R.; He, H.; Matyjaszewski, K. Design and Preparation of Porous Polymers. *Chem. Rev.* **2012**, *112*, 3959-4015.
2. Choi, M.; Ryoo, R. Ordered Nanoporous Polymer-Carbon Composites. *Nat. Mater.* **2003**, *2*, 473-476.
3. Liang, C. D.; Li, Z. J.; Dai, S. Mesoporous Carbon Materials: Synthesis and Modification. *Angew. Chem. Int. Ed.* **2008**, *47*, 3696-3717.
4. Su, D. S.; Delgado, J. J.; Liu, X.; Wang, D.; Schloegl, R.; Wang, L.; Zhang, Z.; Shan, Z.; Xiao, F.-S. Highly Ordered Mesoporous Carbon as Catalyst for Oxidative Dehydrogenation of Ethylbenzene to Styrene. *Chem. Asian J.* **2009**, *4*, 1108-1113.
5. Munoz, B.; Ramila, A.; Perez-Pariente, J.; Diaz, I.; Vallet-Regi, M. MCM-41 Organic Modification as Drug Delivery Rate Regulator. *Chem. Mater.* **2003**, *15*, 500-503.
6. Wu, Z. X.; Zhao, D. Y. Ordered Mesoporous Materials as Adsorbents. *Chem. Commun.* **2011**, *47*, 3332-3338.
7. Zhai, Y.; Dou, Y.; Zhao, D.; Fulvio, P. F.; Mayes, R. T.; Dai, S. Carbon Materials for Chemical Capacitive Energy Storage. *Adv. Mater.* **2011**, *23*, 4828-4850.
8. Zhang, F.; Meng, Y.; Gu, D.; Yan, Y.; Chen, Z.; Tu, B.; Zhao, D. An Aqueous Cooperative Assembly Route to Synthesize Ordered Mesoporous Carbons with Controlled Structures and Morphology. *Chem. Mater.* **2006**, *18*, 5279-5288.
9. Liang, C. D.; Hong, K. L.; Guiochon, G. A.; Mays, J. W.; Dai, S. Synthesis of a large-scale highly ordered porous carbon film by self-assembly of block copolymers. *Angew. Chem. Int. Ed.* **2004**, *43*, 5785-5789.
10. Meng, Y.; Gu, D.; Zhang, F.; Shi, Y.; Yang, H.; Li, Z.; Yu, C.; Tu, B.; Zhao, D. Ordered mesoporous polymers and homologous carbon frameworks: amphiphilic surfactant templating and direct transformation. *Angew. Chem. Int. Ed.* **2005**, *44*, 7053-7059.

Mater. Res. Soc. Symp. Proc. Vol. 1549 © 2013 Materials Research Society
DOI: 10.1557/opl.2013.1056

Study and characterization of the carbon-based nanoparticles obtained from PET

Alena Borisovna Kharissova[1], Edgar de Casas Ortiz[2], Oxana V. Kharissova[2], Ubaldo Ortiz Mendez[1],
Boris I. Kharisov[3]
[1]FIME and CIIDIT, Universidad Autónoma de Nuevo León, Mexico
[2]FCFM and CIIDIT, Universidad Autónoma de Nuevo León, Mexico. E-mail okhariss@mail.ru
[3]FCQ and CIIDIT, Universidad Autónoma de Nuevo León, Mexico

Material like PET {polyethylene terephthalate $(C_{10}H_8O_4)_n$} are usually thrown away present in glasses of refreshments, water bottles between others which are hard to be degraded. However, this material can be recycled and used to acquire nanostructures. During this investigation the objective was to obtain nanoparticles and carbon based nanostructures from the polymer type PET by means of microwave irradiation at the temperature of 260°C at normal pressure and at 600 psi in the presence of acids, ethylene glycol and by means of calcinations. The obtained nanoparticles of ultrananocrystalline diamonds were studied by means of scanning electron microscopy (SEM), high-resolution transmission electron microscopy (TEM), and Raman spectroscopy.

Keywords: PET, nanodiamonds, high-resolution transmission electron microscopy, Raman spectroscopy.

INTRODUCTION

Polymers have been proved to be effective precursors to form different carbon nanostructures.[1] In particular, polyehyleneterephthalate (PET) was used as precursor of microporous carbons,[2] hydrogenated amorphous carbon films[3] and other carbon nanomaterials[4] or for the modification of precursors in carbon materials manufacture.[5] Also, it can be recycled as activated carbons for electrode material in supercapacitors.[6] In these transformations above, classic heating methods and laser decomposition of polymer were applied. At the same time, hydrothermal microwave heating is relatively new technique which is currently being developed very rapidly.[7] Diamond particles of 2–20 nm in size have attracted an increased attention in the past years[1]. Nanodiamond, also called nanocrystalline diamond (NCD) powder, or ultra-dispersed diamond (UDD), is considered a promising material for various applications, including abrasives for the semiconductor and optical industries, extra durable and hard coatings, additives to lubricants for engines and moving gears, polymer reinforcements, protein adsorbents, and even medicinal drugs[1, 5-7]. Nanodiamond powders produced by detonation of explosives in a closed chamber have been commercially available in Russia and Ukraine for over 15 years, and most publications on this topic originate from these countries. In this paper, we describe the application of this method for PET treatment, which led to formation of ultrananocrystalline diamonds.

EXPERIMENTAL

For the experiment, ethylene glycol was used as a reducing agent and nitric acid was used as a strong oxidizing agent in different proportions with PET ($C_{10}H_8O_4)_n$). The reactions were carried out in a Teflon autoclave (equipment MARS-5) at a temperature of 260°C reaching pressure close to 600 psi due to the necessity to exceed boiling points of nitric acid (83°C) and of ethylene glycol (187°C) to acquire a significant degradation in the PET. Some samples were first calcined for 0.5 hour at 600°C in the muffle furnace and then finely grind before proceeding with the reaction in the Teflon autoclave. The reaction times in the MARS-5 were of 10 min at 400 W. The formed nanostructures were studied by Raman spectroscopy (RAMAN equipment Thermo Scientific DRX 532 nm) and high-resolution transmission electron microscopy (HRTEM, equipment Hitachi H-9500).

RESULTS AND DISCUSSION

The samples of PET with ethylene glycol (ratios of 1:1, 1:5, 1:10 without previous calcinations) and PET with ethylene glycol (ratios of 1:1, 1:5 and 1:10 with previous calcinations) were analyzed by high-resolution TEM. Amorphous structures were found in all samples (see Fig. 1a as an example) with a slight degradation but without crystalline nanostructures (Fig. 1b).

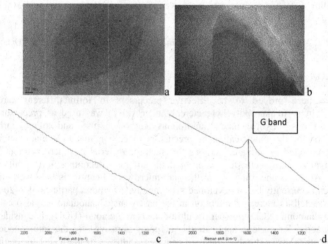

Fig. 1 (a) HRTEM of PET with ethylene glycol without previous calcinations (1:10 ratio) resulting in an amorphous structure; (b) HRTEM of PET with ethylene glycol and with a previous treatment of calcination (1:10 ratio) resulting in an amorphous structure; (c) Raman image of PET with ethylene glycol (c) Raman image of PET with ethylene glycol at 1:10 ratio;(d) Raman image of PET with previous calcinations ethylene glycol and previous calcinations at 1:10 ratio.

Different results were observed in case of PET with ethylene glycol with a previous treatment of calcination. It was found by the SEM-EDS that, in case of calcination treatments, the samples showed presence of carbon and of oxygen. The PET samples of ethylene glycol calcinated (1:5 and

1:10) were analyzed in the Raman Spectroscopy. They showed a slight band at 1580 cm^{-1} indicating partial presence of the up-shifted D band of graphite. HRTEM image of PET with ethylene glycol at 1:10 ratio showed the formation of more definite structures than previous probes with ethylene glycol (Fig.1d).

More significant degradation of PET was noticed with the addition of nitric acid at different concentrations (1:1, 1:5 and 1:10). Fig. 2 (HRTEM image) showes crystal-like nanostructures at the highest proportion of nitric acid. Mixed crystalline and amorphous structures for the nanoparticles produced from this process could be possibly formed. Raman analysis of PET with nitric acid at 1:8 and 1:16 ratios indicated a very similar behavior, having two band shifts at 1332 cm^{-1} and at 1600 cm^{-1}.

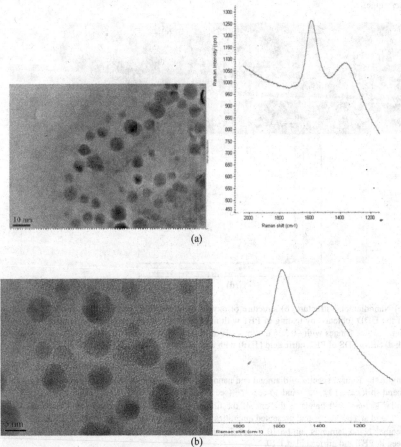

(a)

(b)

Fig. 2. (a) (left) HRTEM image of structure of PET with nitric acid at 1:1 ratio; (right) Raman spectroscopy of PET with nitric acid at 1:1 ratio with band shift of 1332 cm^{-1} and 1600 cm^{-1}; (b) (left) HRTEM image of structure of PET with nitric acid at 1:5 ratio; (right) Raman spectroscopy of PET with nitric acid at 1:5 ratio with band shift of 1332 cm^{-1} and 1600 cm^{-1}.

If weak D and G (or D′) bands of disordered graphite were distinguishable on the as-produced UDD, the sintered sample showed only a smooth background without carbon bands visible. The disorder-induced D and D′ carbon bands are believed to be associated with a double-resonance Raman effect. A strong maximum in the phonon density of states (DOS) of graphite at ~1580 cm^{-1} may also contribute to D′ band for more detailed explanation and discussion of carbon bands). The diamond peak at ~1320 cm^{-1} is downshifted and broadened (FWHM of ~30 cm^{-}) with respect to the single crystal diamond peak (1332 cm^{-1}) (Fig. 2). This downshift is thought to occur due to phonon confinement or changes in the phonon DOS accompanying the decrease of particles size into the nanometer range.

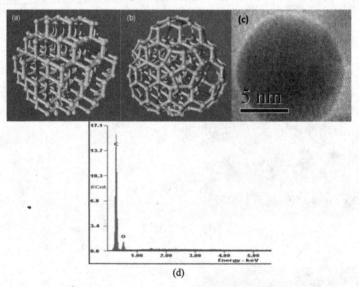

Fig. 3. a) Nanodiamond structure, b) structure of oxygen-functionalied nanodiamond, c) HRTEM image of the UDD obtened by teating of PET with nitric acid (1:10) with a previous treatment of calcination. Lattice fringes with ~0.206 nm inter-planar distances correspond to the (111) planes of diamond; d) SEM-EDS of PET nitric acid (1:10) with a previous treatment of calcination.

According to the Raman spectra of diamond and nanocrystalline diamond,[8] the structures have two Raman band shifts at 1332cm^{-1} and 1618cm^{-1} (see Fig. 2). The Raman spectra of Fig. 2 showed very similar Raman shift bands to the one of the ultracrystalline nanodiamonds having shifts at 1340cm^{-1} and at 1600cm^{-1} and presented quantities of amorphous material, according the Raman spectroscopy.[8] Nanodiamonds can be 10 nm of diameter in nanosized tetrahedral networks which can be seen in PET and nitric acid at 1:10.

CONCLUSIONS

As the main result of this work, it has been established that the degradation of PET by means of nitric acid at 1:10 and 1:5 with the previous treatment of calcinations, grind and Teflon autoclave, allows the formation of ultrananocrystalline diamonds (of size about 10 nm, or less) in agglomerates with a Raman band shift, at 1340 cm^{-1} and 1620 cm^{-1}. Comparing the obtained data with those reported earlier,[9] we observed a close similarity in positions of these Raman bands, which were attributed to sp^3 carbon (diamond) and sp^2 carbon (graphite), respectively. We suggest that the importance of this work consists in the possibility of useful utilization of PET wastes for fabrication of nanodiamonds. As it is well-known, nanodiamond is versatile material for biological applications, such as biosensors, drug carriers, and imaging probes, so its relatively simple and cheap production method could give much opportunities on its further uses.

BIBLIOGRAPHY

[1] Maksimova, N.I.; Krivoruchko, O.P.; Mestl, G.; Zaikovskii, V.I.; Chuvilin, A.L.; Salanov, A.N.; Burgina, E.B. Catalytic synthesis of carbon nanostructures from polymer precursors. *Journal of Molecular Catalysis A: Chemical*, **2000**, *158*, 301–307.

[2] Fernández-Morales, I.; Almazán-Almazán, M.; Pérez-Mendoza, M.; Domingo-García, M.; López-Garzón, F.J. PET as precursor of microporous carbons: preparation and characterization. *Microporous and Mesoporous Materials*, **2005**, *80* (1–3), 107–115.

[3] Budai, J.; Bereznai, M.; Szakacs, G.; Szilagyi, E.; Toth, Z. Preparation of hydrogenated amorphous carbon films from polymers by nano- and femtosecond pulsed laser deposition. *Applied Surface Science*, **2007**, *253* (19), 8235-8241.

[4] Zhuo, C.; Alves, J.O.; Tenorio, J.A.S.; Levendis, Y.A. Synthesis of Carbon Nanomaterials through Up-Cycling Agricultural and Municipal Solid Wastes. *Industrial & Engineering Chemistry Research*, **2012**, *51* (7), 2922-2930.

[5] Barriocanal, C.; Díez, M.A.; Alvarez, R. PET recycling for the modification of precursors in carbon materials manufacture. *Journal of Analytical and Applied Pyrolysis*, **2005**, *73* (1), 45–51.

[6] Centeno, T.A.; Rubiera, F.; Stoeckli, F. Recycling of residues as precursors of carbons for supercapacitors. 1st Spanish National Conference on Advances in Materials Recycling and Eco – Energy, Madrid, 12-13 November 2009, S04-2, pp. 95-98.

[7] Kharisov, B.I.; Kharissova, O.V.; Ortiz Mendez, U. Microwave hydrothermal and solvothermal processing of materials and compounds. In: *"Microwave Heating"*, INTECH, **2012**, Edited by: Wenbin Cao, ISBN 979-953-307-840-2.

[8] Hodkiewicz, J. Characterizing carbon materials with Raman spectroscopy. **2010**, In: Madison, WI, USA: Thermo Fisher Scientific Inc.

[9] Subramanian, K.; Kang, W.P.; Davidson, J.L.; Takalkar, R.S.; Choi, B.K.; Howell, M.; Kerns, D.V. Enhanced electron field emission from micropatterned pyramidal diamond tips incorporating $CH_4/H_2/N_2$ plasma-deposited nanodiamond. Proceedings of Diamond 2005, the 16th European

Conference on Diamond, Diamond-Like Materials, Carbon Nanotubes, Nitrides & Silicon Carbide. *Diamonds & Related Materials*, **2006**, 15(4-8), 1126–1131.

Mater. Res. Soc. Symp. Proc. Vol. 1549 © 2013 Materials Research Society
DOI: 10.1557/opl.2013.843

Diamond Dielectrics for Advanced Wakefield Accelerators

Stanley S. Zuo[1], James E. Butler[1], Bradford B. Pate[2], Sergey P. Antipov[1], Alexei Kanareykin[1], Chunguang Jing[1]

[1] Euclid TechLabs, 5900 Harper Rd, #102, Solon, OH 44139, U.S.A
[2] Naval Research Laboratory, 4555 Overlook Ave. SW, Washington, DC 20375, U.S.A

ABSTRACT

Diamond was investigated as one of the superior dielectric materials for advanced wakefield accelerators. Both planar and cylindrical wakefield accelerating structures were constructed. An AsTex microwave plasma-enhanced CVD system was modified for synthesis of cylindrical polycrystalline diamond tubes. Cylindrical diamond tubes were successfully synthesized from hydrogen and methane and are characterized with micro Raman, photoluminescence spectroscopy and optical tests. In addition, planar wakefield structures were constructed from commercially available diamond. Wakefield tests on a rectangular diamond structure confirm that diamond can sustain microwave electric field strengths of 0.3 GV/m at its surface without material breakdown.

INTRODUCTION

Dielectric loaded accelerating (DLA) structures excited by high current electron beams or an external high frequency, high power, RF source have been under extensive study for many years [1,2,3,4]. In recent years, electromagnetic wakefields produced by high energy electrons transiting through dielectric loaded waveguides have pushed electric fields to GV/m levels in the microwave and mm-wave frequency range [3,4,5,6,7]. Dielectric materials such as metal oxide ceramics and fused silica are encountering their physical limits. Diamond is an attractive material for dielectric loaded accelerating structures [5,8,9,10] due to its low microwave loss tangent at Ka-W frequency bands, excellent thermal conductivity, and high RF breakdown field [2,5]. The basic RF structure is very simple - a cylindrical, dielectric loaded waveguide with an axial vacuum channel is inserted into a conductive sleeve (Figures 1,2). A high charge, (10 – 100 nC), short, (1 – 4 mm) electron drive beam generates TM_{01} mode electromagnetic Cherenkov radiation (wakefields) while propagating down the vacuum channel. Following at a delay adjusted to catch the accelerating phase of the wakefield is a trailing electron (witness) bunch. The witness beam is accelerated to high energy by the wakefield produced by the drive beam.

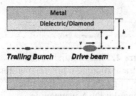

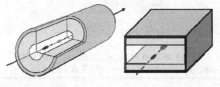

Figure 1. Cross-sectional view of a rectangular (or cylindrical if a and b are radii) dielectric-loaded waveguide.

Figure 2. Design of cylindrical and rectangular diamond DLA structures and their correlated waveguides.

The principal goal of this investigation is to develop a diamond-based DLA to allow a sustained accelerating gradient larger than 600 MV/m, far in excess of the limits experimentally observed for conventional metallic accelerating structures. The key element to achieve this is to obtain diamond material with low impurity content and uniform quality over the finished geometry that can sustain such high fields. The use of microwave plasma-enhanced CVD (MPCVD) was determined to have a greater likelihood of success based on its more rapid deposition rate and the ability to control substrate temperatures during deposition [11,12]. In this investigation, we report on the deposition and characterization of polycrystalline cylindrical diamond tubes for advanced DLA structures, grown by microwave plasma-enhanced chemical vapor deposition.

EXPERIMENT

A 5 kW AsTex microwave plasma-enhanced CVD cylindrical applicator was modified for the synthesis of the cylindrical diamond structure. As in the original design, microwave power propagates down from the top of the applicator to a water-cooled flat circular substrate. The applicator has been modified by including a slowly rotating metal tube (a mandrel for diamond tube deposition) with adjustable tuning structures to confine the plasma around the mandrel. The temperature of the mandrel (i.e. diamond growth temperature) is maintained by cooling an inner tube with chilled water, and by adjusting the Helium gas pressure surrounding the cooling tube inside the mandrel. A simplified overall cross section of the setup is depicted in Figure 3. Typical conditions for tube deposition ranges from 90 to 120 Torr with 1 to 3% of methane (99.999 % purity) in hydrogen at a mandrel temperature of 800 to 1000 °C. The incident power used is between 3 to 4 kW. No additional additive gas is used. After the growth, the diamond tube is released from the mandrel by acid etching. Trimming and final finishing will be accomplished by laser machining.

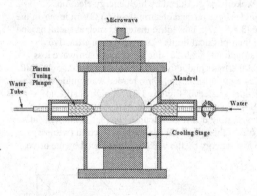

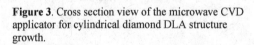

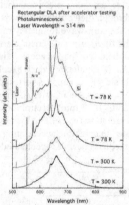

Figure 3. Cross section view of the microwave CVD applicator for cylindrical diamond DLA structure growth.

Figure 4. Photoluminescence spectra at room and low temperature (78 K) of a PCD plate from the rectangular DLA waveguide.

Two types of DLA diamond structures were synthesized and constructed – cylindrical diamond tube structure in cylinder waveguide and planar diamond structure in rectangular waveguide [13]. The actual structure is confined in the metal assembly, built being part of the beam line of the particle accelerator and it is a complex device overall. Simplified drawings of these structures are illustrated in Figure 2. The cylindrical structure is a simple polycrystalline diamond cylinder. The external surface is coated with copper before it is assembled in the copper sleeve. One planar structure is constructed with two parallel single crystal diamond plates. Another planar structure is constructed with two pieces of freestanding polycrystalline CVD diamond wafers. The overall parameters of the DLA structure were determined via parametric simulations with constraint on the thickness of commercially available diamond plates. The minimal beam gap was determined by beam dynamics.

As a proof-of-principle for the survivability of diamond in high intensity wakefields, deep (200 um) and narrow (20 um) grooves were cut in the surface of the SCD diamond plates to enhance the wakefield induced electric fields. Electric fields at least of 0.3 GV/m were present on the diamond surface in the groove [14]. Inspection of the diamond was performed before and after the beam test under scanning electron microscopy. No apparent breakdown damage, such as burn, crack or material vaporization, was observed. Figure 4 shows the Photoluminescence spectra of the PCD plate from the planar DLA diamond structure at both room temperature and low temperature (78 K) after the PCD diamond plates were exposed to high-energy electron beam during the breakdown test. Numerous numbers of points on different segments of the diamond plates were examined. The absence of the neutral vacancy, i.e. the GR1 center, was noted in measured spectra, indicating that displacement damage [15,16,17] is below the detection limit. Excitation was by an Argon ion laser (514.5 nm wavelength).

Table I. Characteristics and test accelerating parameters of the dielectric-loaded waveguide structures. Wakefield gradient is assessed from a single Gaussian beam charge bunch.

Load-in material		SCD plate	PCD plate	PCD tube
Operating frequency	f_0	24.81 GHz	250 GHz[4]	20.8 GHz
Beam channel height/ID	$2a$	4.0 mm	200 μm	3.22 mm
Waveguide height/OD	$2b$	6.4 mm	350 μm	6.825 mm
Waveguide geometry		Rectangular	Rectangular	Cylinder
Waveguide type		Resonator	Travelling wave	Resonator
Waveguide width	w	8.0 mm	1.0 mm	NA
Waveguide length	L	4.0 mm	6.0 cm	40.0 mm
Material thickness	h	1.2 mm	75 μm	1.80 mm
Relative permittivity	ε_r	5.7	5.7	5.7
Loss tangent	σ_d	1×10^{-4}	1×10^{-4}	1×10^{-4}
Beam Characteristics		(2.5 mm, 70 nC)	(0.32 mm,70 pC)	TBD
Wakefield Gradient		1 MV/m/nC	154.3 MV/m/nC	TBD

Wakefield DLA breakdown testing experiments are performed at two accelerator test facilities due to the different structure geometries and dimensions [14]. High energy, high current and short charge bunches is produced from an L-band photocathode RF gun (drive gun) and two standing wave linac structures. Then the produced high velocity moving charge bunches pass through the dielectric-loaded structure and create the Wake where high electric field is therefore present. The DLA geometry and dimension are carefully assessed, calculated and chosen based on the fit to the beam line character, the impedance matching conditions, and the goal of the testing. Multiple bunches of charges (bunch train) can also be used to increase the total energy. In our test experiments, a 2.5 mm long (bunch length) and 70 nano-Coulomb (nC) Gaussian beam charge bunch was used for the single crystal diamond (SCD) planar structure test at Argonne Wakefield Accelerator Testing Facility [14]. A 1.5 mm long square current beam with 500 pico-Coulomb (pC) charge bunch was used for the polycrystalline diamond (PCD) planar structure at Brookhaven Accelerator Testing Facility. The cylindrical polycrystalline diamond tube is planned to be tested with a train of 10 bunches, each 2.5 mm long and 100 nC in size. Table I includes all characteristics and test accelerating parameters of some of the DLA diamond structures tested in our experiments.

DISCUSSION

Figure 5 shows several polycrystalline cylindrical diamond DLA structures synthesized in the modified microwave plasma-enhanced CVD system. The overall length of these tubes is 50 to 60 mm, with a central 30 to 40 mm that is of similar diameter. The inner diameter is 3.2 mm with a wall thickness as thick as 2.5 mm. The bottom image in figure 5 show the transmission of light through a tube (mandrel has yet to be removed).

Figure 6 shows the photoluminescence spectrum of a cylindrical diamond tube sample – NRL5k414. From these spectra, we find that nitrogen (N-V center produces broad feature peaking at ~ 650 nm) and silicon (Si-V center at ~738 nm) are present throughout this diamond tube. The sharp feature at "562 nm" is the diamond Raman line (1332 cm^{-1}). Excitation was by an Argon ion laser (514.5 nm wavelength).

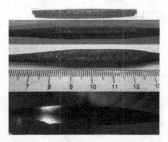

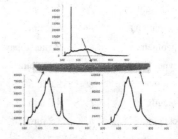

Figure 5. Cylindrical diamond DLA tubes synthesized in the modified microwave plasma-enhanced CVD system. The lower three diamond tubes are still on the mandrel.

Figure 6. Photoluminescence spectrum of a diamond DLA tube – sample NRL5k414.

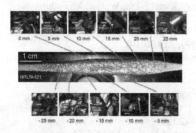

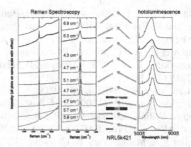

Figure 7. Optical microscopy of a cylindrical diamond DLA tube – sample NRL5k421.

Figure 8. Raman, FWHM values and photo-luminescence spectroscopy of a cylindrical diamond DLA tube – sample NRL5k421.

Figure 7 shows the microscopy of a cylindrical diamond tube sample – NRL5k421. Optical microscopy indicates crystalline uniformity along the central section of the tube, with signs of lower quality diamond (small crystallite size and/or dendritic features) towards the end of the tube. Figure 8 shows both the Raman and photoluminescence spectra of a cylindrical diamond tube sample – NRL5k421. The Raman line FWHM indicate uniform quality diamond deposited for significant length of the tube. Excitation was by an Argon ion laser (514.5 nm wavelength).

SUMMARY

We have reported our motivation, proof of principle accelerator experiments, fabrication techniques (water cooled rotating mandrel in a commercial microwave plasma CVD system) and the current status of our ongoing development of diamond materials for DLA structures. Optical microscopy, and Raman and Photoluminescence spectroscopy characterization are presented that find tube quality and dimensional uniformity over a 4 cm length was achieved. Proof-of-principle accelerator tests have been carried out for rectangular diamond DLA structures that confirm diamond can sustain exposure to electric field strengths of 0.3 GV/m at its surface without material break down. The synthesized diamond tubes reported on here will be fabricated by laser machining into cylindrical wakefield structures for testing under high field gradients of the upgraded Argonne Wakefield Accelerator Test Facility.

ACKNOWLEDGMENTS

This work is supported by DOE SBIR grant DE-FG02-08ER85033 (Euclid Techlabs LLC).

REFERENCES

1 W. Gai, P. Schoessow, B. Cole, R. Konecny, J. Norem, J. Rosenzweig, and J. Simpson, Phys. Rev. Lett. 61, 2756 (1988).
2 M. C. Thompson, H. Badakov, A. M. Cook, J. B. Rosenzweig, R. Tikhoplav, G. Travish, I. Blumenfeld, M. J. Hogan, R. Ischebeck, N. Kirby, R. Siemann, D. Walz, P. Muggli, A. Scott, and R. B. Yoder, Phys. Rev. Lett. 100, 214801 (2008).
3 A. Kanareykin. Journal of Physics, Conf. Ser., 236, 012032 (2010).

4   S. Antipov, C. Jing, P. Schoessow, S. Zuo, J. Butler, A. Kanareykin, W. Gai, M. Fedurin, K. Kusche, and V. Yakimenko, IPAC'12, Conf. Proc., 2816 (2012).
5   A. Kanareykin, AIP Conference Proceedings, v. 1299, 286 (2010).
6   S. Antipov et al., AIP Conference Proceedings, v. 1299, 359 (2010).
7   A. Kanareykin, J.E. Butler, S. P. Antipov, P. Schoessow, C. Jing, S. Zuo, IPAC'12, Conf. Proc., 2819 (2012).
8   D. Whittum, SLAC-PUB-7910, July 1998.
9   A. Kanareykin, et al., EPAC 2006 Conf. Proc., 2460 (2006).
10 S. Antipov, C. Jing, A. Kanareykin, P. Schoessow, M. Conde, W. Gai, S. Doran, J.G. Power, Z. Yusof, AIP Conf. Proc, PAC'11, New York, 2074 (2011).
11 J.E. Butler Y.A. Mankelevich, A. Cheesman, Jie Ma and M.N. R. Ashfold, J. Phys.: Condens. Matter 21, 364201 (2009).
12 S.S. Zuo, M.K. Yaran, T.A. Grotjohn, D.K. Reinhard, J. Asmussen, Diamond Relat. Mater. 17, 300 (2008).
13 S. Antipov, C. Jing, A. Kanareykin, J. E. Butler, V. Yakimenko, M. Fedurin, K. Kusche, and W. Gai, Appl. Phys. Lett. 100, 132910 (2012).
14 S. Antipov et al, IPAC'12, Conf. Proc., 2813 (2012).
15 C.D. Clark and C.B. Dickerson, Philosophical Transactions: Physical Sciences and Engineering, Vol. 342, No. 1664, Thin Film Diamond, 253 (1993).
16 A. Wotherspoon, J.W. Steeds, P. Coleman, D. Wolverson, J. Davies, S. Lawson, J.Butler, Diamond Relat. Mater. 11, 692 (2002).
17 D. Fisher, D.J.F. Evans, C. Glover, C.J. Kelly, M.J. Sheehy, G.C. Summerton, Diamond Relat. Mater. 15, 1636 (2006).

AUTHOR INDEX

SUBJECT INDEX

Printed in the United States
By Bookmasters